U0084842

湯圓糯米糰

變化62種甜品

大福、芝麻球、菓子，

教你變化花樣多變的吃法！

小三 著

湯圓 糯米糰 變化62種甜品

大福、芝麻球、菓子，
教你變化花樣多變的吃法！

作　　者　小三

發 行 人　程安琪
總 策 畫　程顯灝

總 編 輯　呂增娣
主　　編　李瓊絲、鍾若琦
特約編輯　呂增慧
資深編輯　程郁庭
編　　輯　許雅眉、鄭婷尹
編輯助理　陳思穎
美術總監　潘大智
資深美編　劉旻旻
美　　編　游騰緯、李怡君
行銷企劃　謝儀方、吳孟蓉

發 行 部　侯莉莉
財 務 部　許麗娟
印　　務　許丁財
出 版 者　橘子文化事業有限公司

總 代 理　三友圖書有限公司
地　　址　106 台北市安和路 2 段 213 號 4 樓
電　　話　(02) 2377-4155
傳　　真　(02) 2377-4355
E－mail　service@sanyau.com.tw
郵政劃撥　05844889 三友圖書有限公司

總 經 銷　大和書報圖書股份有限公司
地　　址　新北市新莊區五工五路 2 號
電　　話　(02) 8990-2588
傳　　真　(02) 2299-7900

製　　版　興旺彩色印刷製版有限公司
印　　刷　鴻海科技印刷股份有限公司
初　　版　2015 年 7 月
定　　價　新臺幣 300 元
ＩＳＢＮ　978-986-364-049-3 (平裝)

國家圖書館出版品預行編目 (CIP) 資料

湯圓．糯米糰變化 62 種甜品 - 大福、芝麻球、菓子，教你變化花樣多變的吃法！ / 小三作 .
-- 初版 . -- 臺北市：橘子文化，2015.06
面；　公分
ISBN 978-986-364-049-3(平裝)

1. 點心食譜

427.16　　104009518

序・幸福滋味

從沒想到有機會為別人的著作寫序，所以當小三邀請我為她的新書寫序的時候，我感到非常榮幸，也替她高興。

美食是人們生活的一部份，烹調是不分國界，身為專業人員，應該要透過雙手做出能讓人感動的料理，要懂得將美食與生活合二為一，這才是一種樂趣。

這本書將為讀者示範了 60 多道湯圓及糯米糰子美食，只要跟著書上的內容逐步製作便能做出好食的湯圓，讀者們透過巧手更能將它們變成一道道具有生命及帶給人幸福的美食。

令人感動的美食對我們而言，是烹調者在過程中用心設計、細心烹調，將愛心傳達所做出的製成品，自己動手做美食，讓自己及所愛的人享用那份幸福滋味。

姚滿輝 師傅

香港中式飲食業資歷架構（四級）評核員

香港中式飲食業技能提升計劃培訓導師

國家專業資格中式麵點高級技師

國家職業技能鑒定麵點考評員

序・為湯圓注入新元素

屈指一算，認識小三原來已有 8 年時間，大家結緣於「美味 DIY」網路討論區，每當我們遇到做甜品的難題，無論成功與否，大家也會在網上一起分享心得。及後，在大型聚會上真正認識她，除了談談烹飪經外，也會聊聊生活逸事，慢慢成為好友。

小三為人單純率真，具有責任感，她的「烏龍」事跡亦多籮籮，有她參與的聚會，席上絕無冷場，為大家帶來不少歡笑。她多年來跟從名師學習，累積不少入廚心得，對烹飪充滿熱誠。

此書以中國傳統食品「湯圓」為題，除了揉合不同地方的精髓外，更為「湯圓」注入新元素，令它可以經歷時代變遷，令大眾對湯圓的一貫看法刮目相看。

近來，得知她生活上有點小波折，但看見她慢慢將興趣化成工作，更有機會將多年來的烹飪心得集結成書，實在替她感到高興。在此，祝願她事事如意，一紙風行！

小肥糖 Candy Tsui
（人氣 Blogger）

前言

傳統習俗中，許多人會在冬至及喜慶的日子煮湯圓來食用。湯圓象徵著一家團圓的意思，每吃下一口湯圓，就像擁有了一份圓滿的祝福。時至今日，湯圓已成為日常食品，很多人下午茶或宵夜時都會煮湯圓來吃。

從前人們要吃湯圓，可得費盡功夫！首先是磨米製粉造湯圓皮。那是將糯米浸泡一夜，然後用石磨磨成米漿，當米漿流入棉布袋內，要用繩將袋口綁緊，在上面放石磨，好將水分擠出，形成濕粉。然後做湯圓餡，包湯圓和煮湯圓……工夫一大串，所以只有特別的日子才會吃。

相較於現代人而言，想吃湯圓實在方便多了，湯圓可塑性也高，可大可小，有鹹有甜，顏色豐富，有餡沒餡都各具特色，不論你是大朋友或小孩子，這一本書相信總有你喜歡的湯圓。

團圓美滿，是每一個人心目中最大的幸福，而湯圓對我來說更加包括我對爸爸的思念，因為爸爸喜歡吃湯圓……

這本書裏製作的 60 多款湯圓，既有傳統的、家鄉的食製，也有利用各地風味湯羹配製的創新口味，還有東洋風格的乾糯米糰子。希望大家做得順手，一家大小吃得開心！

小三

目錄

BEFORE
第一次
做湯圓

PART 1
沁心透涼的糯米糰子

PART 2
絲絲暖意的甜湯圓

PART 3 家鄉風味的鹹湯圓

PART 4 自家餡料變出新花款

BEFORE

第一次做湯圓

多變化的糯米食材——湯圓與糯米糰子

湯圓象徵圓滿、團圓，許多人更喜愛它的口感，因此在不同地區、不同族群都有不同的演繹。

廣式湯圓

糖水湯底大多數是用薑汁或薑片煮片糖而成，而餡料有豆沙、芝麻，但以薑汁片糖湯圓為最經典，是最傳統的廣式湯圓，在昔日物資短缺的年代，母親都會做給兒女們吃的甜品，內餡是包入一小粒片糖，那份愛如片糖一般甜蜜。

上海湯圓

有鹹有甜，鹹湯圓的餡料有雞肉、豬肉、冬菇，而甜湯圓的餡料有芝麻泥、玫瑰豆沙和蓮泥等，皮薄餡多。鹹甜湯圓的湯底大多數比較清淡。上海道地名點酒釀丸子，以甜酒釀為湯，加入糯米小丸子，深具特色。

客家湯圓

有鹹有甜，鹹的湯圓顏色有紅有白，形狀呈圓圓小小一粒一粒，沒有包餡，湯底比較豐富，用雞湯、肉絲、蝦米、冬菇、茼蒿、芹菜、油葱…等；而甜的就將糯米粉蒸熟，然後剪成一小顆的麻糬，食用時沾上花生糖粉。

如何搓湯圓

原味湯圓皮

1 糯米粉放入大碗中，加入適量清水

2 混合成糰

3 搓揉成軟滑的長條

4 切粒

5 搓圓成皮料

6 可包餡或不包餡

包餡湯圓

1 皮料及餡料分別搓圓

2 皮料用食指開窩

3 放入餡料

4 左手大拇指輕按餡料，右手用虎口收緊皮料

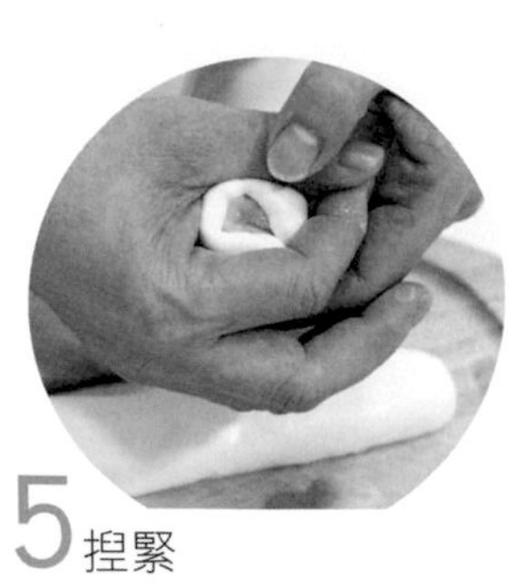

5 捏緊

6 收口

混合其他材料的湯圓皮

1 番薯蒸熟

2 趁熱壓成泥

3 加入糯米粉內混合

4 加入適量清水

5 混合

6 成糰

7 搓揉成軟滑的長條，切粒

8 搓圓成皮料

TIPS: 要視情況加清水

◎要按混合入粉糰的材料本身所含水量，調節應加入的水量多寡。基本上搓好粉糰的軟硬度應好像我們的耳珠般。

◎番薯蒸熟後含水量很低，要加入多一些水分調節，而南瓜蒸熟後會有很多水分，可能一點水也不用加。

湯圓煮法須知

如何煮湯圓

1 煮滾半鍋清水，放入湯圓

2 煮至浮起及脹身

3 撈起

4 泡冰水

煮湯圓的重點叮嚀

搓湯圓時需要加入多少水分才足夠？

因為當中存在太多變數，很難將糯米粉與水做一個黃金比例，不同牌子的糯米粉吸水能力各有不同，當加入南瓜、番薯之類作皮料就更加難掌握水的分量，因為每種蔬果本身的含水量分別很大，所以書中搓糯米糰的水分只供參考，新手最好是將水分逐次加入，搓揉至如耳珠軟硬度即可，如果粉糰太乾可以加入少許清水，相反如果太濕可以加入少許糯米粉，只要多做幾次，必定能找出自己喜歡的水比例。

應該使用什麼水搓揉糯米糰？

- 用什麼溫度的水都可以，只是做出來的粉糰所做的湯圓口感略有不同。
- 涼水或冰水搓揉，簡單直接，操作容易，但並不耐煮。
- 溫水或熱水搓揉，可以增加粉糰的黏性，包餡時比較不易裂開，延展性較佳，耐煮不易破裂，吃時比較有彈性。但熱水太多會令粉糰會太黏，容易黏手，反而不易操作。
- 先熱後冷，先加入部分熱水拌勻，再加入冷水搓揉，互補不足。
- 冷熱各有優劣，任君選擇。

湯圓要煮多久才熟？

- 水滾下湯圓，煮至浮起，再煮至漲身，因為湯圓煮至內裏的水氣向外才會脹大，亦即表示內裏已經熟透。
- 如果湯圓體積大或是餡料比較難熟，可以在湯圓煮滾時加入適量冷水再煮滾至漲身，即可保證熟透。
- 不要煮得過久，以免糯米皮爛掉。

水分充足才能煮好湯圓嗎？

煮湯圓一定要夠多水，以便翻滾時有足夠空間，不時要用湯匙推動湯圓附近的水，以免它們互相黏貼在一起。

湯圓也要「過冷河」？

將煮好的湯圓泡冰水才放入熱糖水裏，可以保持外型美觀，增加進食時湯圓的彈性。

湯圓皮黏手，怎麼辦？

包湯圓時，雙手可以沾少許油，會較易操作。

我煮的湯圓，常常漏餡，怎麼辦？

包湯圓時，湯圓皮不宜太薄，還要與餡料緊貼，否則湯圓煮熟膨脹，皮就會被熱空氣逼得裂開，漏出餡料。

怎樣才可令湯圓包得貼服呢？

包湯圓時，湯圓皮不要像包子那樣圍在餡料外捏，要在粉糰開洞，將餡料塞入去，並把餡料與湯圓皮之間的空氣擠出去，才能包貼。

湯圓可以直接放在湯或糖水中煮嗎？

湯圓不要貪懶一鍋煮。不管煮甜湯或鹹湯，煮湯圓都需要另外起一鍋水煮熟，這樣湯圓才不會煮得過於軟爛及湯頭也不至於太混濁。

為什麼湯圓不能當早餐？

湯圓因為黏性高，不易消化，我們早起時體內腸胃功能是最弱的，最好不要在早餐時進食湯圓。

TIPS: 糖水的多寡

煮湯圓的糖水，甜度因人而異，本書所建議的分量只供參考。

PART 1
沁心透涼的糯米糰子

大福草莓
Strawberry Daifuku

數量：7 個 / pieces
烹製時間：30 分鐘 / mins

材料

糯米粉 140 克
清水 250 克
砂糖 1 湯匙
植物油 1 茶匙
草莓 7 粒
豆沙餡 150 克
糕粉 適量

Ingredients

140g glutinous rice flour
250g water
1 tbsp sugar
1 tsp oil
7 pcs strawberry
150g red bean paste
Some cooked glutinous rice flour

做法

1. 草莓洗淨，去蒂，用廚房紙抹掉水分。
2. 豆沙餡分 7 等份，每份豆沙包裹 1 粒草莓，弄圓，備用。
3. 糯米粉與砂糖混合，加入植物油、清水拌勻成稀漿，將稀漿倒入已塗油的盤子內，蒸約 15 ～ 20 分鐘。
4. 大盆內放入熟糕粉作手粉，將蒸熟的粉糰放入糕粉內，調整粉糰，分割 7 份，每份包入一份餡料，草莓底部向上收口，即可享用。

Methods

1. Wash strawberries, remove stalk, and pat dry with kitchen paper.
2. Divide red bean paste into 7 portions. Wrap in 1 pc of strawberry, roll to round shape, and set aside.
3. Mix glutinous rice flour and sugar. Add in oil and water, mix well to form dilute mixture. Pour onto plate which is spread with oil, and steam for 15 ～ 20 mins.
4. Pour cooked soybean powder onto large tray, and put steamed dough into soybean powder. Roll dough, divide into 7 portions, wrap in filling (with the bottom of strawberry faces upwards), seal the opening and serve.

番薯芋圓

Sweet Potato and Taro Balls

數量：4～6人份 / serves
烹製時間：60分鐘 / mins

材料

湯圓料

・番薯圓

番薯 200克（去皮切片）
糯米粉 120克
清水 適量

・芋圓

芋泥 150克
糯米粉 120克
溫水 適量

湯料

番薯 200克（去皮切粒）
薑 20克（刨皮拍扁）
片糖 適量

Ingredients

Sweet potato balls

200g sweet potato (peeled and sliced)
120g glutinous rice flour
Some water

Taro balls

150g taro puree
120g glutinous rice flour
Some water

Sweet potato soup

200g sweet potato (peeled and diced)
20g ginger (peeled and grated)
Some lump sugar

做法

1 切片番薯蒸熟，趁熱壓泥加入糯米粉混合，加入清水搓揉成軟滑粉糰，切粒，搓圓。
2 芋泥蒸熱，加入糯米粉混合，加入清水搓揉成軟滑粉糰，切粒，搓圓。
3 清水加薑塊煮滾，加入番薯粒煮至滾起後下片糖煮至糖溶解。
4 另煮滾半鍋清水，放入芋圓、番薯圓煮至浮起及脹身，撈起，將湯圓放糖水中享用。

TIPS:

片糖可以黃砂糖代替。

Methods

1 Steam sliced sweet potato, mash to puree when it is hot. Add in glutinous rice flour and water, mix to form smooth dough. Divide and roll to round shape.
2 Steam taro puree. Add in glutinous rice flour and water, mix to form smooth dough. Divide and roll to round shape.
3 Boil water with ginger. Add in diced sweet potato. When boiled, add in lump sugar, and boil until sugar is dissolved.
4 Boil half pot of water. Add in taro and sweet potato balls, boil until floating and swell, and drain. Add into sweet soup, and serve when hot.

翡翠明珠

Pandan Rice Balls in Coconut Sweet Soup

數量：4 ～ 6 人份 / serves
烹製時間：30 分鐘 / mins

材料

湯料

椰漿 400 克
牛奶 300 克
玉米粒 150 克
砂糖 100 克
清水 400 克

湯圓料

· 斑蘭湯圓

斑蘭葉 3 ～ 4 片
糯米粉 150 克
清水 120 克
綠豆餡 200 克

Ingredients

Sweet soup

400g coconut milk
300g milk
150g corn
100g sugar
400g water

Pandan rice balls

3 ～ 4 pcs pandan leaf
150g glutinous rice flour
120g water
200g green bean paste

做法

1 綠豆泥分 20 等份，搓圓，備用。
2 清水煮滾，加入玉米粒煮滾，加入椰漿及牛奶拌勻，最後放入砂糖拌至糖溶，備用。
3 斑蘭葉清洗後剪成小段（約 1 寸），放入攪拌機中，加入適量清水打成斑蘭液。
4 糯米粉放入大碗中，加入斑蘭液及清水搓揉成粉糰，再分切成 20 等份，搓圓，用食指開窩，包入番薯餡，收口捏緊，搓圓。
5 燒滾半鍋清水，放入湯圓煮至浮起及脹身，撈出湯圓，放入玉米椰漿內，趁熱享用。

Methods

1 Divide green bean paste into 20 portions, roll round and set aside.
2 Boil water and add in corn. When boiled, add in coconut milk and milk, mix well. Add sugar and stir until sugar is dissolved. Set aside.
3 Wash pandan leaf and cut into small piece (about 1 inch). Put into blender, add in water, and blend to form pandan juice.
4 Pour glutinous rice flour into large bowl. Add in pandan juice and water, mix to form smooth dough. Divide into 20 portions, and roll to round shape. Use forefinger to make a hole, wrap in green bean paste, seal up the opening and roll into round shape.
5 Boil half pot of water. Add in rice balls, boil until floating and swell, and drain. Add into corn and coconut milk sweet soup, and serve when hot.

色彩繽紛小丸子

Colorful Rice Balls

數量：2 ～ 4 人份 / serves

烹製時間：40 分鐘 / mins

材料

湯圓料

- 斑蘭湯圓
  - 斑蘭葉 2 片
  - 糯米粉 90 克
  - 清水 80 克
- 紅色小丸子
  - 糯米粉 90 克
  - 食用紅色素 少許
  - 清水 80 克
- 幻紫小丸子
  - 芋泥粉糰 85 克
  - 紫米糰 85 克

湯料

- 涼粉 100 克
- 芒果布丁 100 克
- 椰纖果 100 克

Ingredients

100g grass jelly
100g mango pudding
100g nata de coco

Pandan rice balls
2 pcs pandan leaf
90g glutinous rice flour
80g water

Red small rice balls
90g glutinous rice flour
1 drop red edible coloring
80g water

Purple small rice balls
85g taro dough
85g purple rice dough

做法

1. 涼粉沖冰開水，切粒。
2. 糯米粉加入斑蘭液及清水搓揉成粉糰，切粒，搓圓。
3. 紅色小丸子糯米粉加入紅色素，加清水搓揉成軟滑的粉糰，再分切成小丸子，搓圓。
4. 淺紫及深紫粉糰分別搓成長條，然後將兩條扭在一起，切粒，搓圓。
5. 燒滾半鍋清水，放入湯圓煮至浮起及脹身，撈出湯圓，攤冷後可以放入冰箱，隨時可以享用，進食時加入糖漿，即可。

Methods

1. Rinse grass jelly with cold drinking water and dice.
2. Mix glutinous rice flour with pandan juice and water. Mix to form dough, divide and roll to round shape.
3. Add red edible coloring and water into glutinous rice flour. Mix to form smooth dough, divide and roll to round shape.
4. Roll light purple and dark purple dough separately into elongated shape. Twist them together, divide and roll to round shape.
5. Boil half pot of water. Add in rice balls, boil until floating and swell, and drain. Let cool and can be kept in refrigerator. Serve with syrup.

四式湯圓綠豆仁

Four Colored Rice Balls in Green Bean Sweet Soup

數量：4 ～ 6 人份 / serves

烹製時間：60 分鐘 / mins

材料

湯圓料

・芋泥湯圓
- 熱芋泥 45 克
- 糯米粉 45 克
- 清水 80 克

・番薯湯圓
- 熱番薯泥 45 克
- 糯米粉 45 克
- 清水 80 克

・紫米湯圓
- 糯米粉 90 克
- 紫米水 80 克

・原味湯圓
- 糯米粉 90 克
- 清水 80 克

湯料
- 綠豆仁 150 克
- 冰糖 適量

Ingredients

150g skinless green bean
Some crystal sugar

Original rice balls
90g glutinous rice flour
80g water

Taro puree rice balls
45g hot taro puree
45g glutinous rice flour
80g water

Purple rice balls
90g glutinous rice flour
80g purple rice water

Sweet potato rice balls
45g hot sweet potato puree
45g glutinous rice flour
80g water

做法

1 綠豆仁沖洗後，用清水浸泡 2 小時，瀝乾水分，加入清水煮熟。
2 芋泥蒸熱，加入糯米粉混合，加入清水搓揉成軟滑粉糰，切粒，搓圓。
3 糯米粉加入紫米水，搓揉成軟滑的長條，切粒，搓圓。
4 番薯蒸熟，趁熱壓泥加入糯米粉混合，加入清水搓揉成軟滑粉糰，切粒，搓圓。
5 糯米粉加入清水搓揉成軟滑粉糰，切粒，搓圓。
6 加入全部材料煮滾，趁熱進食。
7 綠豆仁煮滾加入冰糖拌至糖溶解，加入其餘湯圓一起進食。

Methods

1 Wash green bean and soak in water for 2 hrs, and drain. Add in water and boil until well done.
2 Steam taro puree, add in glutinous rice flour and mix well. Add in water and roll to form smooth dough. Divide and roll to round shape.
3 Mix glutinous rice flour and purple rice water. Roll to form smooth and elongated dough. Divide and roll to round shape.
4 Steam sweet potato, mash to puree and add in glutinous rice flour when hot. Add in water and roll to form smooth dough. Divide and roll to round shape.
5 Mix glutinous rice flour and water. Roll to form smooth dough. Divide and roll to round shape.
6 Boil all ingredients together. Serve when hot.
7 Boil skinless green bean to well done, add in crystal sugar and stir until sugar is dissolved. Add in rice balls and serve.

蜜紅豆雙色冰淇淋

Sweet Red Bean Dual-color Ice-cream

數量：1 ～ 2人份 / serves
烹製時間：20 分鐘 / mins

材料

冰淇淋 2 球
清水 95 克
蜜紅豆 適量

湯圓料

・番薯粒與小丸子

熱番薯 120 克
糯米粉 150 克
食用紅色素 少許

Ingredients

2 scoops of ice-cream
95g water
Some sweetened red bean

120g hot sweet potato
150g glutinous rice flour
Little red edible coloring

做法

1 番薯蒸熟半份切粒，另外半份趁熱加入糯米粉 50 克，搓揉成軟滑的粉糰，切粒，搓圓。
2 餘下糯米粉用清水拌勻，然後分兩份，一份原色，另一份加入紅色素，分別搓揉成軟滑的粉糰，再分切成小丸子，搓圓。
3 燒滾半鍋清水，放入全部小丸子煮至熟，撈起，放入冰水內降溫。
4 將小丸子、番薯粒及蜜紅豆放入碗內，再放入冰淇淋，即可享用。

Methods

1 Steam sweet potato. Dice 1/2 portion of sweet potato and the remaining portion mix with 50g glutinous rice flour when hot. Mix to form smooth dough, divide and roll to round shape.
2 Mix the remaining glutinous rice flour with water. Divide into 2 portions. One is kept in original color while another portion add with red edible coloring. Handle separately to form smooth dough. Divide and roll to round shape.
3 Boil half pot of water. Add in all rice balls, boil until floating and swell, and drain. Add into icy water to lower temperature.
4 Put small rice balls, diced sweet potato and sweetened red bean into bowl. Place 2 scoops of ice-cream on top. Serve.

醉酒葡萄

Colour Rice Balls with Wine Grape

數量：**2 ～ 3** 人份 / serves

烹製時間：**40** 分鐘 / mins

材料

湯圓料

・斑蘭小丸子
斑蘭葉 2 片
糯米粉 75 克
清水 70 克

・紅色小丸子
糯米粉 75 克
食用紅色素 少許
清水 70 克

湯料

・琴酒葡萄乾
青葡萄乾 8 ～ 12 粒
琴酒（氈酒） 4 湯匙
芒果 1 個

Ingredients

Pandan rice balls
2 pcs pandan leaf
75g glutinous rice flour
70g water

Red small rice balls
75g glutinous rice flour
Little red edible coloring
70g water

Gin-soaked raisins
8 ～ 12 golden raisin
4 tbsp gin
1 pc mango

做法

1 提早一晚將青葡萄乾浸在琴酒中。
2 斑蘭葉清洗後剪成小段(約 1 寸)，放入攪拌機中，加入適量清水打成斑蘭汁。
3 糯米粉放入大碗中，加入斑蘭汁及清水，搓揉成粉糰，再分切成小丸子，搓圓成綠色粉圓。
4 清水裏加入 1 滴食用紅色素，拌糯米粉搓揉成軟滑的粉糰，切成小丸子，搓圓。
5 燒滾半鍋清水，放入兩種小丸子煮至熟，撈起，放入冰水內降溫。
6 芒果去皮取果肉，將果肉放入攪拌機內加入少許冰塊打成芒果汁。
7 兩色小丸子和琴酒葡萄乾放入玻璃杯內，然後淋上新鮮芒果汁進食。冷凍後享用，風味更佳。

Methods

1 Soak raisins in gin one night before.
2 Wash pandan leaf and cut into small piece (about 1 inch). Put into mixer, add in water, and mix to form pandan juice.
3 Pour glutinous rice flour into large bowl. Add in pandan juice and water, mix well to form smooth dough. Divide into little pieces and roll to round shape.
4 Drop red edible coloring into water and add into glutinous rice flour, mix well to form smooth dough. Divide into little pieces and knead to round shape.
5 Boil half pot of water. Add in both rice balls, boil until floating and swell. Drain and put into icy water to cool.
6 Peel mango and remove stone. Put mango flesh into blender, add ice, blend as mango juice.
7 Put rice balls and gin-soaked raisins into glass, pour mango juice on top. It may serve chilly.

紫薯南瓜盅

Purple Sweet Potato and Pumpkin Cup

數量：10 個 / pieces

烹製時間：20 分鐘 / mins

材料

南瓜泥 60 克

紫薯餡 150 克

糯米粉 60 克

再來米粉（粘米粉） 10 克

砂糖 10 克

Ingredients

60g pumpkin puree

150g purple sweet potato paste

60g glutinous rice flour

10g rice flour

10g sugar

做法

1 紫薯餡分成 10 等份，分別搓圓，放入冰箱冷凍定型，備用。

2 南瓜蒸熟趁熱壓成泥狀，將南瓜泥、糯米粉、再來米粉及砂糖一起放入大碗中混合，搓揉成軟滑有光澤的長條，分切 10 等份，搓圓。

3 用食指開窩，放入紫薯餡，用虎口收緊皮料至紫薯餡約 2/3 位置即可。

4 將南瓜盅菓子放入已經鋪了蒸紙或掃油的蒸鍋內，大火蒸約 4 ～ 6 分鐘，即可。

Methods

1 Divide purple sweet potato paste into 10 portions. Roll to round shape and store in refrigerator.

2 Steam pumpkin, and mash to puree when it is hot. Pour pumpkin puree, glutinous rice flour, rice flour and sugar into large bowl, mix to form smooth and elongated dough. Divide into 10 portions and roll to round shape.

3 Use forefinger to make a hole, wrap in purple sweet potato filling, tighten the wrap at two third of height.

4 Put pumpkin cup onto steamer with steamed sheet or spread with oil. Steam at high heat for 4 ～ 6 mins and serve.

TIPS:

南瓜泥如果是預先製作，一定要蒸熟才可以加入粉內搓揉，搓揉時如果粉糰太乾，可以加入少許清水，相反如果太濕可以加入少許糯米粉。

Pumpkin puree should be prepared before use. You must steam pumpkin until well done before adding flour and rolling. If dough is too dry, you may add little water. If dough is too wet, you may add little glutinous rice flour.

湯圓串

A string of Rice balls

數量：**20** 串 / pieces

烹製時間：**30** 分鐘 / mins

TIPS:

加入溫水搓揉，加強湯圓的彈性。

沾料也可以改為用蜜糖加溫水少許調開再加炒香的黑芝麻。

Add in warm water to roll so as to increase the elasticity of rice balls.

You may mix little warm water with honey, and add in stir-fried black sesame to replace peanut candy.

材料

湯圓料

・棕色湯圓

巧克力粉 2 茶匙
糯米粉 130 克
砂糖 20 克
溫水 110 克

・原味湯圓

糯米粉 130 克
砂糖 20 克
溫水 110 克

・紫色湯圓

蒸熟紫番薯 50 克
糯米粉 70 克
砂糖 20 克
清水 100 克（視情況加減）

伴食料及工具

熟黃豆粉 1 碗
黃砂糖 1 碗
竹籤 20 支

做法

1 巧克力湯圓料拌勻，搓揉成軟滑的長條，切成 20 等份，搓圓。
2 紫番薯趁熱壓泥加入糯米粉、砂糖混合，加入清水搓揉成軟滑的長條，切成 20 等份，搓圓。
3 原味湯圓料拌勻，搓揉成軟滑的長條，切成 20 等份，搓圓。
4 熟黃豆粉加黃砂糖拌勻成黃豆糖。
5 燒滾半鍋清水，放入湯圓煮至浮起及脹身，撈出湯圓，放入冰水中降溫，撈起瀝乾。
6 竹籤沾水，每支串入 3 顆不同顏色的湯圓，沾上黃豆糖，即可享用。

Ingredients

Brown rice balls

2 tsp cocoa powder
130g glutinous rice flour
20g sugar
110g warm water

Purple rice balls

50g steamed purple sweet potato
70g glutinous rice flour
20g sugar
100g water
(adding amount is subject to actual situation)

Original rice balls

130g glutinous rice flour
20g sugar
110g warm water

Topping

1 bowl cooked soybean powder
1 bowl lump sugar
20 bamboo sticks

Methods

1 Mix cocoa rice balls ingredients. Roll to form smooth and elongated dough. Divide into 20 portions and roll to round shape.
2 Steam purple sweet potato, mash to puree when it is hot. Add in glutinous rice flour, sugar and water, and roll to form smooth and elongated dough. Divide into 20 portions and roll to round shape.
3 Mix original rice balls ingredients. Roll to form smooth and elongated dough. Divide into 20 portions and roll to round shape.
4 Mix cooked soybean powder with lump sugar.
5 Boil half pot of water. Add in rice balls, boil until floating and swell, remove and drain. Place into icy water to lower temperature, then drain.
6 Dip bamboo stick with water. Prick one each of rice balls with different flavours. Dip soybean sugar, and serve.

芒果布丁小丸子

Rice Balls with Mango Pudding

數量：2 ～ 3人份 / serves

烹製時間：60 分鐘 / mins

材料

芒果果凍粉 1 盒
熱水 114 克
冰水 142 克
煉乳 114 克
芒果 1 個（取果肉，切丁）

湯圓料

・紅色小丸子
糯米粉 90 克
食用紅色素 少許
清水 80 克

・原味湯圓
糯米粉 90 克
清水 80 克

Ingredients

1 box mango jelly powder
114g hot water
142g iced water
114g evaporated milk
1 pc mango (peel, remove pit and dice)

Red rice balls
90g glutinous rice flou
1 drop red edible coloring
80g water

Original rice balls
90g glutinous rice flour
80g water

做法

1 將果凍粉用熱水拌至溶解，然後加入冰水拌勻， 當溫度降低後，才拌入煉乳成布丁液。
2 先將少量布丁液加入芒果丁內，然後將其餘布丁液倒入容器內，稍後，將芒果丁連布丁液一起倒入剛才的容器內拌勻，芒果丁便不會沉底。
3 糯米粉加入抹茶及溫水搓揉成軟滑粉糰，切粒，搓圓。
4 糯米粉加入紫米水搓揉成軟滑粉糰，切粒，搓圓。
5 糯米粉加入清水搓揉成軟滑粉糰，切粒，搓圓。
6 布丁倒扣在盤上，然後排上其他小丸子，可以伴以蜜紅豆進食，風味更佳。

Methods

1 Dissolve jelly powder in hot water. Add in iced water and mix well. When temperature is lowered, add in evaporated milk to form pudding mixture.
2 Add in little pudding mixture to mango. Pour the remaining pudding mixture into container. After a while, pour mango with pudding liquid mixture into the container, and mix well. In that way, mango will not sink.
3 Mix glutinous rice flour, green tea and warm water. Mix well to form smooth dough, divide and roll to round shape.
4 Mix glutinous rice flour with purple rice water. Roll well to form smooth dough, divide and roll to round shape.
5 Mix glutinous rice flour with water. Mix well to form smooth dough, divide and roll to round shape.
6 Turn mango pudding upside down, and place pudding onto plate. Arrange small rice balls on the plate, and serve with sweetened red beans.

西米南瓜露斑蘭丸子

Pandan Rice Balls with Sago and Pumpkin Sweet Soup

數量：4 ～ 6 人份 / serves

烹製時間：40 分鐘 / mins

材料

湯料

南瓜 300 克
（去皮取果肉）
椰漿 150 克
牛奶 250 克
清水 600 克
砂糖 80 克
西谷米 40 克
清水 適量

湯圓料

・斑蘭湯圓

斑蘭葉 3 ～ 4 片
糯米粉 150 克
芋泥餡 200 克
清水 120 克

Ingredients

Sweet soup

300g pumpkin (peeled)
150g coconut milk
250g milk
600g water
80g sugar
40g sago
Some water

Pandan rice balls

3 ～ 4 pcs pandan leaf
150g glutinous rice flour
200g taro paste
120g water

做法

1. 南瓜切片蒸熟，趁熱壓成泥；芋泥餡分成 20 等份，備用。
2. 西谷米泡水 15 分鐘，瀝乾，加入適量清水煮至中心有一點白，關火焖至白點消失，沖水。
3. 清水煮滾，加入南瓜泥、椰漿及牛奶拌勻，加入砂糖拌溶；最後放入西谷米，備用。
4. 斑蘭葉清洗後剪成小段（約 1 寸），放入攪拌機中，加入適量清水打成斑蘭液。
5. 糯米粉放入大碗中，加入斑蘭液及清水搓揉成粉糰，再分切成 20 等份，搓圓，用食指開窩，包入芋泥餡，收口捏緊，搓圓。
6. 燒滾半鍋清水，放入湯圓煮至浮起及漲身，撈出湯圓，放入南瓜糊內，趁熱享用。

Methods

1. Steam pumpkin, slice and mash into puree when it is hot. Divide taro paste into 20 portions. Set aside.
2. Soak sago in water for 15 mins, and drain. Add in water and boil until there is little white spot at the center. Turn heat off, and braise until white spot disappears. Rinse under running water to cool.
3. Boil water, add in pumpkin puree, coconut milk and milk, mix well. Add in sugar and stir until sugar is dissolved. Add in sago at last.
4. Wash pandan leaf and cut into small piece (about 1 inch). Put into blender, add in water, and blend to form pandan juice.
5. Pour glutinous rice flour into large bowl. Add in pandan juice and water, roll to form smooth dough. Divide into 20 portions, and roll to round shape. Use forefinger to make a hole, wrap in taro paste, seal up the opening and roll into round shape.
6. Boil half pot of water. Add in rice balls, boil until floating and swell, and drain. Add into sweet pumpkin soup, and serve when hot.

TIPS:

西谷米在食用前才放入南瓜糊內，口感更加 Q。
Sago is put into sweet pumpkin soup before serve to give better taste of sago.

紅豆抹茶湯圓

Green Tea Rice Balls in Red Bean Sweet Soup

數量：4 ～ 6 人份 / serves

烹製時間：90 分鐘 / mins

材料

湯料

紅豆 300 克
蓮子 50 克
乾百合 76 克
冰糖 70 克
清水 適量

湯圓料

糯米粉 110 克
抹茶粉 1 湯匙
溫水 90 克

做法

1 紅豆沖洗後用清水浸泡 3 小時，濾掉水分；蓮子洗淨，用牙籤將蓮子芯推出；乾百合洗淨。

2 紅豆、蓮子與百合一起放入清水中煮滾，然後調中小火煮約 1 小時，熄火燜 15 分鐘，再開火煮至滾熟，加入冰糖拌至糖溶解。

3 糯米粉與抹茶粉放入碗內，加入溫水搓揉成軟滑粉糰，切粒，搓圓。

4 另煮滾半鍋清水，放入湯圓煮至浮起及脹身，撈起，將湯圓放糖水中享用。

Ingredients

Red bean sweet soup

300g red bean
50g lotus seed
76g dried lily
70g crystal sugar
Some water

Rice balls

110g glutinous rice flour
1 tbsp green tea powder
90g warm water

Methods

1 Wash red bean and dried lily, soak in water for 3 hrs and drain. Wash lotus seed and use tooth pick to push the core out.

2 Put red bean, lotus seed and dried lily in water and bring to a boil. When boiled, turn to medium low heat and cook for about 1 hr. Turn heat off, and braise for 15 mins. Turn heat on again and boil until sticky. Add in crystal sugar and stir until sugar is dissolved.

3 Put glutinous rice flour and green tea powder into bowl. Add in warm water and mix to form smooth dough. Divide and roll to round shape.

4 Boil half pot of water. Add in rice balls, boil until floating and swell, and drain. Add rice balls into sweet soup, and serve.

玲瓏南瓜菓子

Pumpkin Sweet

數量：**15** 個 / pieces

烹製時間：**30** 分鐘 / mins

材料

蒸熟南瓜 120 克
糯米粉 90 克
再來米粉（粘米粉） 25 克
砂糖 10 克
植物油 10 克
芋泥餡 150 克
紅莓乾 15 粒

工具

牛骨籤或竹籤

做法

1 芋泥餡分成 15 等份，分別搓圓，放入冰箱冷凍定型，備用。
2 糯米粉、再來米粉及砂糖一起放入大碗中混合，備用。
3 南瓜蒸熟趁熱壓成泥狀，將南瓜泥加入大碗中拌勻，加入植物油搓揉成軟滑有光澤的長條，分切 15 等份，搓圓，用食指開窩，放入芋泥餡，收口向下再搓圓，稍微按扁。
4 用竹籤壓線塑造南瓜型，最後在頂端插入紅莓做瓜蒂。
5 將南瓜菓子排放入已經鋪了蒸紙或掃油的蒸鍋內，大火蒸約 6 ～ 8 分鐘，即可。

Ingredients

120g cooked pumpkin
90g glutinous rice flour
25g rice flour
10g sugar
10g oil
150g taro paste
15pcs dried cranberry

Equipment

Cow bone pick or bamboo stick

Methods

1 Divide taro paste into 15 portions, roll to round shape and store in refrigerator.
2 Put glutinous rice flour, rice flour and sugar into large bowl, and mix well.
3 Steam pumpkin and mash to puree when it is hot. Put pumpkin puree into large bowl, add in oil, stir well to form smooth and elongated dough. Divide into 15 portions, and roll to round shape. Use forefinger to make a hole, wrap in taro paste, seal up the opening and roll into round shape. Slightly press flat.
4 Use bamboo stick to make pumpkin pattern. Insert 1 cranberry each on top as pumpkin stalk.
5 Arrange sweet pumpkin onto steamer with steamed sheet or spread with oil, and steam over high heat for 6 ～ 8 mins.

TIPS:

南瓜含水量高，搓揉時如果粉糰太乾可以加少許南瓜泥或清水，相反如果太濕可以加少許再來米粉或糯米粉。
加入植物油可令湯圓入口更滑及在造型時不會太黏手。
Pumpkin contains large amount of water. You may add some pumpkin puree or water into glutinous rice flour if too dry whereas add glutinous rice flour or rice flour if too wet.
Add oil to make rice balls smoother and not too sticky when making shape.

番薯芝麻球

Deep-fried Sweet Potato Sesame Ball

數量：16 個 / pieces

烹製時間：30 分鐘 / mins

材料

番薯 50 克（去皮）
糯米粉 150 克
泡打粉 1/4 茶匙
蘇打粉 1/4 茶匙
砂糖 50 克
豆沙餡 240 克
熱水 150 克
清水 適量
白芝麻 1 杯

Ingredients

50g sweet potato (peeled)
150g glutinous rice flour
1/4 tsp baking powder
1/4 tsp baking soda
50g sugar
240g red bean paste
150g hot water
Some water
1 cup white sesame

做法

1 豆沙餡分切 16 等份，搓圓，備用。
2 糯米粉、泡打粉、蘇打粉、砂糖一起放入大碗內混合。
3 番薯切片蒸熟，用湯匙壓成泥。將番薯泥及熱水趁熱一起加入大碗內，並用木棍快速拌勻，之後按情況加入適量清水，用手搓揉成糰，再分切大約 16 等份。
4 包入餡料，收口捏緊並搓成圓形，在表面全部分別沾上白芝麻。
5 準備半鍋油，開火就要立即放入芝麻球，用中火加熱約 3 分鐘，其間不可翻動芝麻球。
6 當炸至變黃就可以用鍋鏟搓磨芝麻球，待脹身及外皮酥脆且顏色變金黃後便可取出，瀝乾油分。

Methods

1 Divide red bean paste into 16 potions. Roll round and set aside.
2 Put glutinous rice flour, baking powder, baking soda and sugar into large bowl. Mix well.
3 Slice sweet potato and steam. Use spoon to mash as puree when hot. Put sweet potato puree and hot water into large bowl. Use wooden stick to mix well quickly. Add in appropriate amount of water, roll to form dough, and divide into 16 portions.
4 Wrap in filling, seal up the opening and roll to round shape. Dip white sesame on surface.
5 Boil half pot of oil. Turn heat on, add in sesame ball immediately, and deep fry at medium heat for 3 mins. Do not turn sesame ball upside down.
6 When in yellow color, use spatula to turn sesame balls. When swell and become crispy with golden brown color, take out and drain.

擂沙湯圓
Rice Balls with Soybean Powder

數量：20 個 / pieces
烹製時間：20 分鐘 / mins

材料

糯米粉 130 克
再來米粉（粘米粉） 20 克
植物油 2 茶匙
清水 140 克
黑芝麻泥 200 克
熟黃豆粉 200 克
糖粉 3 湯匙

Ingredients

130g glutinous rice flour
20g rice flour
2 tsp vegetable oil
140g water
200g black sesame puree
200g cooked soybean powder
3 tbsp icing sugar

做法

1 黑芝麻泥分為 20 等份，搓圓，蓋上保鮮膜，再放入冰箱冷凍定型，備用。
2 糯米粉、再來米粉及植物油拌勻後，加入清水拌勻，搓揉成軟滑的長條，分別搓圓並用食指開窩，包入芝麻泥，收口捏緊，搓圓。
3 燒滾半鍋清水，放入湯圓煮至浮起及脹身，撈出湯圓。
4 熟黃豆粉及糖粉拌勻，將湯圓沾滿黃豆糖粉，便可享用。

Methods

1 Divide black sesame puree into 20 potions. Roll round, cover with plastic wrap, and store in refrigerator.
2 Mix glutinous rice flour, rice flour and vegetable oil. Then add in water, stir well to form smooth and elongated dough. Divide and roll to round shape. Use forefinger to make a hole, wrap in sesame puree, seal up the opening and roll into round shape.
3 Boil half pot of water. Add in rice balls, boil until floating and swell, and drain.
4 Mix cooked soybean powder and icing sugar. Dip rice balls with soybean powder mixture, and serve.

TIPS:

擂沙湯圓要趁熱吃。
加入植物油可令湯圓入口更滑。
Serve sesame paste rice balls when hot.
Adding oil to make rice balls smoother.

椰泥糯米糰

Stuffing Mochi with Shredded Coconut

數量：15 個 / pieces
烹製時間：35 分鐘 / mins

材料

糯米粉 1 杯
砂糖 1/3 杯
椰漿 1/2 杯
清水 1/2 杯
花生醬餡 150 克
椰絲 適量

Ingredients

1 cup glutinous rice flour
1/3 cup sugar
1/2 cup coconut milk
1/2 cup water
150g peanut butter paste
Some shredded coconut

做法

1. 花生醬餡分 15 等份，搓圓，放入冰箱定型，備用。
2. 糯米粉加入砂糖拌勻；加入椰漿拌勻，慢慢加入清水調成稀漿。
3. 清水慢慢加入糖粉內調成稀漿，將稀漿倒入已塗油的盤子內，蒸約 20 分至熟。
4. 椰絲鋪在盆內，將熟粉糰倒入椰絲上，分成 15 份，每份包入一份餡料，搓圓，外皮沾滿椰絲。

Methods

1. Divide peanut butter paste into 15 portions. Roll round and store in refrigerator.
2. Mix glutinous rice flour and sugar. Mix oil and water.
3. Slowly add water into sugar flour to make light batter. Pour onto plate spread with oil, and steam for 20 mins until well done.
4. Spread shredded coconut on tray, add steamed rice dough, and divide the dough into 15 portions. Wrap in filing, roll round, and dip with shredded coconut.

糖不甩

Peanut Coated Sweet Glutinous Rice Ball

數量：4 ～ 6 個 / pieces
烹製時間：30 分鐘 / mins

材料

糯米粉 1 杯
砂糖 2/3 杯
椰漿 1/2 杯
清水 1/2 杯
食油 1 茶匙

沾料

炒香花生碎 1 杯
炒香白芝麻 1/2 杯
黃糖粉 1 杯

Ingredients

1 cup glutinous rice flour
2/3 cup sugar
1/2 cup coconut milk
1/2 glass of water
1 tsp oil

Dipping ingredients

1 cup stir-fried chopped peanut
1/2 cup stir-fried white sesame
1 cup yellow powdered sugar

做法

1 沾料放入大碗內拌勻。
2 糯米粉加入砂糖拌勻；食油、椰漿與清水混合。
3 椰漿水慢慢加入糖粉內調成稀漿，將稀漿倒入已塗油的盤子內，蒸約 20 分鐘至熟。
4 用剪刀將粉糰剪開一大塊，放入大碗內再剪成一粒粒，即可。

Methods

1 Put dipping ingredients into large bowl.
2 Mix glutinous rice flour with sugar. Mix oil, coconut milk and water.
3 Slowly add coconut water into sugar flour to form dilute liquid. Pour onto plate spread with oil, and steam for 20 mins until well done.
4 Use scissors to cut dough into large piece. Cut into small pieces in bowl and serve.

椰絲綠豆糰子

Coconut Rice Balls with Mengbean

數量：4～6個 / pieces
烹製時間：30 分鐘 / mins

材料

糯米粉 130 克
砂糖 120 克
椰漿 1/2 杯
清水 1/2 杯
綠豆仁 250 克
椰絲 適量
食鹽 少許

Ingredients

130g glutinous rice flour
120g cup sugar
1/2 cup coconut milk
1/2 cup water
250g skinless mengbean
some shredded coconut
salt

做法

1. 綠豆仁沖洗後用清水浸泡 1 小時，瀝乾水分後，隔水蒸至剛熟軟硬適中，備用。
2. 椰絲加入少許食鹽拌勻。
3. 糯米粉加入砂糖拌勻；分別加入椰漿及清水，混合調成稀漿。
4. 將稀漿倒入已塗油的盤子內，蒸約 20 分至熟。
5. 糯米糰用湯匙弄成一球球，將糯米球放在綠豆仁中，沾滿外皮全是綠豆仁，即可。
6. 進食前加入椰絲。綠豆糰子可鹹可甜，想吃甜就在加入椰絲後再加砂糖一起進食。

Methods

1. Wash skinless mengbean and soak for 1 hr. Drain. Steam until soft. Set aside.
2. Add a little salt to shredded coconut, mix well.
3. Mix glutinous flour with sugar, add in coconut milk and water, stir well to form dilute batter.
4. Pour the batter onto a greased plate and steam for 20 mins until cooked.
5. Scoop the steamed dough with tablespoon. Place rice balls into mengbean, roll to dip beans as coating.
6. Add shredded coconut when serving. This rice ball can be served as dessert or savory snack. If like it in sweet taste, you may add more sugar on top of shredded coconut.

TIPS:

椰絲加入少許食鹽增香，風味更佳。
Add salt to shredded coconut will enhance the flavour.

PART 2

絲絲暖意的甜湯圓

桂花芝麻湯圓

Preserved Petals of Sweet Osmanthus and Sesame Rice balls

數量：6 ～ 8 人份 / serves
烹製時間：20 分鐘 / mins

材料

糯米粉 160 克
熱水 130 克
黑芝麻泥 200 克
桂花糖 1 湯匙
清水 適量

Ingredients

160g glutinous rice flour
130g hot water
200g black sesame paste
1 tbsp preserved petals of sweet osmanthus (candies)
Some water

做法

1. 黑芝麻泥分為 20 等份，搓圓，備用。
2. 糯米粉放在大碗中，沖入熱水，立刻用湯匙攪拌，當溫度稍微下降，用手搓揉成軟滑有光澤的長條。
3. 再分切成 20 等份，分別搓圓，用食指開窩，包入芝麻餡，收口捏緊，搓圓。
4. 燒滾半鍋清水，放入湯圓煮至浮起及脹身，撈出湯圓，盛入碗中，加入桂花糖伴吃。

Methods

1. Divide black sesame paste into 20 portions, roll to round shape and set aside.
2. Put glutinous rice flour into large bowl, add in hot water, use spoon to stir immediately. When temperature is lowered, use hands to mix into smooth and elongated shape.
3. Divide dough into 20 portions, and roll into round shape. Use forefinger to make a hole, wrap with black sesame paste, seal up the opening and roll into round shape.
4. Boil half pot of water. Add in rice balls, boil until floating and swell, and drain. Put rice balls into bowl, and serve with some preserved petals of sweet osmanthus (candies).

TIPS:

如果芝麻餡太鬆散，可以搓圓後蓋上保鮮膜，再放入冰箱冷凍定型，會容易包餡。

If black sesame paste is too loose, you may roll it to round shape and cover with plastic / food wrap. Then store in refrigerator to make it set. In that way, you can wrap sesame paste more easily.

桂花酒釀丸子

Rice Balls in Fermented Rice Wine with Osmanthus Flowers

數量：4～6人份 / serves

烹製時間：20分鐘 / mins

材料

湯料（糖水）

- 酒釀 1 湯匙
- 桂花糖 1 湯匙
- 冰糖 適量
- 清水 適量（800 ～ 1000 克）

湯圓料

・小丸子

- 糯米粉 200 克
- 食用紅色素 少許
- 清水 190 克

Ingredients

Preserved petals of sweet osmanthus soup

1 tsp wine
1 tbsp preserved petals of sweet osmanthus (candies)
Some crystal sugar
800 ～ 1000g water

Rice balls

200g glutinous rice flour
Some red edible coloring
190g water

做法

1. 糯米粉用部分清水拌勻，然後分兩份，一份原色，另一份加入紅色素，分別搓揉成軟滑的粉糰，再分切成小丸子，搓圓，備用。
2. 燒滾半鍋清水，放入小丸子煮至熟，撈起。
3. 清水煮滾，加入酒釀、桂花糖及冰糖拌至糖溶解。
4. 將適量丸子放入碗內，加入桂花糖水，即可享用。

Methods

1. Mix glutinous rice flour with water, and divide into two portions. One is kept in original color while the other portion is added with red edible coloring. Roll separately to form smooth dough. Divide, roll to round shape, and set aside.
2. Boil half pot of water. Add in rice balls and boil until well done. Drain.
3. Boil water, add in wine, preserved petals of sweet osmanthus (candies) and crystal sugar. Stir until sugar is dissolved.
4. Put some rice balls into bowl, add in preserved petals of sweet osmanthus soup, and serve.

TIPS:

酒釀及桂花糖可以按個人的喜好而加減份量。
食用色素可以使用紅麴粉替代，但顏色會比較深，或是只做原色。
桂花糖水可以用馬蹄粉加清水勾芡變成桂花露。
The quantity of fermented rice wine or osmanthus flowers subject to different person's favorite.
You may use red yeast powder to replace red edible coloring but the resultant color will be darker. Or you may make rice balls in original color only.
You may mix water chestnut powder with water to make thickening as preserved petals of sweet osmanthus soup.

紫米露湯圓

Rice Balls with Purple Rice Sweet Soup

數量：4～6人份 / serves

烹製時間：60 分鐘 / mins

材料

湯料

- 紫米 120 克
- 西谷米 30 克
- 玉米 30 克
- 清水 適量
- 冰糖 80 克

湯圓料

・紫米湯圓

- 糯米粉 90 克
- 紫米水 80 克

Ingredients

120g purple rice
30g sago
30g corn
appropriate amount of water
80g crystal suga

Purple rice balls

90g glutinous rice flour
80g purple rice water

做法

1. 紫米用清水浸泡 2 小時，瀝乾，紫米水可作搓湯圓之用。
2. 西谷米泡水 15 分鐘，瀝乾，加入適量清水煮至中心有一點白，關火焗至白點消失，沖水。
3. 紫米加清水煮滾後調至中小火煮至米熟，加入冰糖煮至糖溶解，加入玉米和西谷米拌勻。
4. 糯米粉放在大碗中，加入紫米水，搓揉成軟滑的長條，分切成小丸子，搓圓。
5. 燒滾半鍋清水，放入湯圓煮至浮起及脹身，撈出湯圓，放入紫米露中，趁熱享用。

Methods

1. Soak purple rice in water for 2 hrs, and drain. Keep the soaked water to make rice ball dough.
2. Soak sago in water for 15 mins, and drain. Add in water, and boil until its centre is pretty white. Turn heat off, braise until the white spot at the centre disappears. Wash with water.
3. Put purple rice into water and boil. When boiled, turn to medium low heat to cook until rice is sticky and well done. Add in crystal sugar and boil until sugar is dissolved. Add in corn and sago, and stir well.
4. Put glutinous rice flour into large bowl, add in purple rice water to form smooth and elongated shape. Divide dough into small portions, and roll into round shape.
5. Boil half pot of water. Add in rice balls, boil until floating and swell, and drain. Add rice balls into purple rice sweet soup, and serve.

番薯豆沙湯圓

Red Bean Rice Balls in Sweet Potato Soup

數量：4～6人份 / serves

烹製時間：60 分鐘 / mins

材料

湯料

黃心番薯 200 克
片糖 1.5 片
薑 3～4 片
清水 800～1000 克

湯圓料

糯米粉 115 克
清水 100 克
豆沙餡 150 克

Ingredients

Sweet soup

200g yellow sweet potato
1.5 pcs lump sugar
3～4 pcs ginger
800～1000g water

Rice balls

115g glutinous rice flour
100g water
150g red bean paste

做法

1 豆沙餡分為 15 等份，搓圓，備用。
2 黃心番薯去皮切方塊狀。
3 清水一鍋，下薑片煮滾，加入番薯煮滾後轉中小火煮到番薯熟，備用。
4 糯米粉加清水拌勻，搓揉成軟滑的長條，分切成 15 等份，分別搓圓，用食指開窩，包入豆沙餡，收口捏緊，搓圓。
5 燒滾半鍋清水，放入湯圓煮至浮起及脹身，撈出湯圓，盛入番薯糖水中。

TIPS:
片糖可以黃砂糖代替。

Methods

1 Divide red bean paste into 15 portions, roll to round shape and set aside.
2 Peel yellow sweet potato and dice.
3 Bring water to aboil. Add in sliced ginger, then sweet potato. When boiled, turn to medium low heat. Cook until sweet potato is well done. Set aside.
4 Put glutinous rice flour into large bowl, add in water and mix well. Mix to form smooth and elongated dough. Divide dough into 15 portions, and roll into round shape. Use forefinger to make a hole, wrap in red bean paste, seal up the opening and roll into round shape.
5 Boil half pot of water. Add in rice balls, boil until floating and swell, and drain. Place into sweet potato soup, and serve.

木瓜銀耳紅棗湯圓

Rice balls in Papaya, White Fungus and Red Date Sweet Soup

數量：4 ～ 6 人份 / serves
烹製時間：60 分鐘 / mins

材料

湯料

- 木瓜 1 個
- 銀耳（白木耳） 18 克
- 紅棗 8 粒
- 桂圓 16 粒
- 清水 適量
- 冰糖 適量

湯圓料

・小丸子

- 糯米粉 90 克
- 食用紅色素 少許
- 清水 80 克

Ingredients

Sweet soup

- 1 papaya
- 18g white fungus
- 8 pcs red date
- 16g dried longan fruit
- Some water
- Some crystal sugar

Rice balls

- 90g glutinous rice flour
- 1 drop red edible coloring
- 80g water

做法

1. 將木瓜刨皮，去籽，切塊；將白木耳泡軟，切塊；將紅棗泡軟，去核。
2. 煮滾適量清水，放入糖水材料煮滾後調小火煮 1 ～ 1.5 小時，加入冰糖煮至糖溶解。
3. 糯米粉用部分清水拌勻，然後分兩份，一份原色，另一份加入紅色素，分別搓揉成軟滑的粉糰，再分切成小丸子，搓圓。
4. 燒滾半鍋清水，放入小丸子煮至熟，撈起，加入糖水中。

Methods

1. Peel papaya, remove seed and dice. Soak white fungus and slice. Soak red date and remove pit.
2. Boil water and add in sweet soup ingredients. When boiled, turn to low heat and cook for 1 ～ 1.5 hrs. Add in crystal sugar and boil until sugar is dissolved.
3. Mix glutinous rice flour with water. Divide into two portions. One is kept in original color. Add red edible coloring into another portion. Mix separately to form smooth and elongated dough, divide and roll to round shape.
4. Boil half pot of water. Add in rice balls, boil until well done, and drain. Add into sweet soup, and serve.

踏雪尋梅
Snow Red Mochi

數量：4～6人份 / serves
烹製時間：30分鐘 / mins

材料

糯米粉 140 克
砂糖 90 克
椰漿 1/2 杯
清水 1/2 杯
植物油 1 茶匙
紫薯紅莓餡 130 克
椰絲 適量

Ingredients

140g glutinous rice flour
90g sugar
1/2 cup coconut milk
1/2 cup water
1 tsp oil
130g purple sweet potato and cranberry filling
Some shredded coconut

做法

1. 紫薯紅莓餡分成 10 等份，分別搓圓，放入冰箱冷凍定型，備用。
2. 糯米粉加入砂糖拌勻；油、椰漿與清水混合。
3. 將椰漿水加入糖粉調成稀漿，將稀漿倒入已塗油的盤子內，蒸約 15～20分鐘。
4. 使用電鍋的塑膠飯匙兩支輔助，將蒸熟的粉糰搓出一小糰，放入餡料，搓成球狀，再沾上椰絲，即可享用。

Methods

1. Divide sweet potato and cranberry filling into 10 portions. Roll to round shape and store in refrigerator.
2. Mix glutinous rice flour with sugar. Mix oil, coconut milk and water.
3. Add sugar and flour mixture into coconut milk mixture. Pour onto plate (spread oil onto the plate before use) and steam for about 15～20 mins.
4. Use 2 rice ladles as support on both sides, cut the steamed dough to small portions. Wrap in filling, roll to round shape, dip with shredded coconut and serve.

TIPS:

蒸起的粉糰非常黏，操作時可以戴上膠手套。
Steamed dough is very sticky. You may wear plastic gloves during operation.

鳳梨玫瑰湯圓
Pineapple Rose Rice balls

數量：4 ～ 6 人份 / serves
烹製時間：20 分鐘 / mins

材料

糯米粉 150 克
清水 120 克
鳳梨餡 180 克
玫瑰花果醬 適量
冰水 適量

Ingredients

150g glutinous rice flour
120g water
180g pineapple filling
Some roselle jam
Some icy water

做法

1 鳳梨餡分為 18 等份，搓圓；玫瑰花果醬放入擠花袋內，備用。
2 糯米粉放在大碗中，慢慢加入清水，搓揉成軟滑的長條。
3 再分切成 18 等份，分別搓圓，用食指開窩，包入鳳梨餡，收口捏緊，搓圓。
4 燒滾半鍋清水，放入湯圓煮至浮起及脹身，撈出湯圓，放入冰水中降溫。
5 湯圓降溫後瀝乾水分，放在盤子上，擠上玫瑰花果醬，即可。

Methods

1 Divide pineapple filling into 18 portions and roll to round shape. Put roselle jam into piping bag, and set aside.
2 Put glutinous rice flour into large bowl, add in water slowly, and stir to form smooth and elongated dough.
3 Divide into 18 portions, and roll to round shape. Use forefinger to make a hole, wrap in pineapple filling, seal up the opening and roll into round shape.
4 Boil half pot of water. Add in rice balls, boil until floating and swell, and drain. Add into icy water to lower temperature.
5 After cool, drain, and put onto plate. Pipe roselle jam onto rice balls, and serve.

TIPS:

可以按個人的喜好，用其他果醬替代玫瑰花果醬。
You may use jam in any flavourite instead of roselle jam as topping.

紅棗薑汁湯圓
Rice Balls in Red Date Ginger Soup

數量：3 ～ 4 人份 / serves
烹製時間：30 分鐘 / mins

材料

湯料（糖水）

紅棗 6 粒
片糖 100 克
薑汁 1 湯匙
清水 適量

湯圓料

糯米粉 120 克
清水 110 克
片糖 1 條（餡）

Ingredients

Sweet soup

6 pcs red date
100g lump sugar
1 tbsp ginger juice
water

Rice balls

120g glutinous rice flour
110g water
1 pc lump sugar (as filling)

做法

1. 片糖 1 條，用刀切粒，再挑選 15 粒大小適中的，備用。
2. 紅棗泡軟，去核，加清水煮滾，加入薑汁、片糖煮成糖水，備用。
3. 糯米粉放在大碗中，慢慢加入清水，搓揉成軟滑有光澤的長條。
4. 再分切成 15 等份，每份包入一粒片糖，搓圓。
5. 燒滾半鍋清水，放入湯圓煮至浮起及脹身，撈出湯圓，放入紅棗薑汁糖水內，即可享用。

Methods

1. Dice 1 pc of lump sugar. Select 15 pcs in proper size, and set aside.
2. Soak red date to make it soft. Remove the pit from red date, and boil in water. Add in ginger juice and lump sugar to make it as sweet soup. Set aside.
3. Put glutinous rice flour into large bowl, add in water slowly, and roll to form smooth and elongated shape.
4. Divide into 15 portions, wrap in 1 dice of lump sugar, and roll to round shape.
5. Boil half pot of water. Add in rice balls, boil until floating and swell, and drain. Add into red date ginger sweet soup, and serve.

TIPS:

片糖可以黃砂糖代替。

百年好合湯圓

Rice Balls in Lotus Seed and Dried Lily Soup

數量：*6* ～ *8* 人份 / serves

烹製時間：*60* 分鐘 / mins

材料

湯料（糖水）

蓮子 100 克
乾百合 40 克
桂圓 25 粒
紅棗 15 粒
冰糖 適量
清水 1200 克

湯圓料

糯米粉 150 克
清水 140 克
豆沙餡 200 克

Ingredients

Sweet soup

100g lotus seed
40g dried lily
25 pcs dried longan fruit
15 pcs red date
Some crystal sugar
1200g water

Rice balls

150g glutinous rice flour
140g water
200g red bean paste

做法

1. 豆沙餡分 18 等份，搓圓，放入冰箱定型，備用。
2. 蓮子洗淨，用牙籤將蓮子芯推出；紅棗泡軟，去核。
3. 煮滾一小鍋水，放入蓮子煮至再滾起，轉小火煮至軟身，撈起，瀝乾水分。
4. 百合洗淨，放入清水中，用中火煮約 20 分鐘，加入蓮子、紅棗煮片刻，加入冰糖拌至糖溶解，備用。
5. 糯米粉放在大碗中，慢慢加入清水，搓揉成軟滑的長條，分切成 18 等份，分別搓圓，用食指開窩，包豆沙餡，收口捏緊，搓圓。
6. 燒滾半鍋清水，放入湯圓煮至浮起及脹身，撈起，將湯圓放入糖水內，即可享用。

Methods

1. Divide red bean paste into 18 portions, roll to round shape, and store in refrigerator.
2. Wash lotus seed and use tooth stick to push lotus seed pod out. Soak red date to make it soft, and remove the pit from the red date.
3. Boil a pot of water, add in lotus seed and boil. When boiled, turn to low heat and boil until soft. Remove and drain.
4. Wash dried lily. Boil in water over medium heat for about 20 mins. Add in lotus seed and red date, boil for a while. Then add in crystal sugar until dissolved. Set aside.
5. Put glutinous rice flour into large bowl, add in water slowly, mix to form smooth and elongated shape. Divide dough into 18 portions, and roll into round shape. Use forefinger to make a hole, wrap with some red bean paste, seal up the opening and roll into round shape.
6. Boil half pot of water. Add in rice balls, boil until floating and swell, and drain. Add rice balls into sweet soup, and serve.

四喜養生湯圓

Rice Balls in Healthy Sweet Soup

數量：4 ～ 6 人份 / serves
烹製時間：60 分鐘 / mins

材料

湯料

銀耳（白木耳） 1 大個
紅棗 12 粒
桂圓 12 粒
枸杞 10 克
冰糖 適量
清水 1200 克

湯圓料

豆沙餡 150 克
糯米粉 120 克
清水 110 克

Ingredients

Sweet soup

1 large pc white fungus
12 pcs red date
12 pcs dried longan fruit
10g medlar fruit
some crystal sugar
1200g water

Rice balls

150g red bean paste
120g glutinous rice flour
110g water

做法

1 銀耳泡軟，切小塊；桂圓、枸杞沖洗，瀝乾。
2 紅棗泡軟，去核，加清水煮滾，調中火後加入銀耳、桂圓、枸杞煮約 30 分鐘，加入冰糖拌至糖溶解。
3 糯米粉放在大碗中，慢慢加入清水，搓揉成軟滑有光澤的長條，再分切成 15 等份，用食指開窩，包入豆沙餡，收口捏緊，搓圓。
4 燒滾半鍋清水，放入湯圓煮至浮起及脹身，撈出湯圓，放入糖水內，即可享用。

Methods

1 Soak white fungus to make soft, and chop into small pieces. Wash dried longan fruit and medlar fruit, and drain.
2 Soak red date to make soft. Remove pit from the red date, and boil in water. When boiled, turn to medium heat, and add in white fungus, dried longan fruit and medlar fruit, cook for about 30 mins. Add in crystal sugar and stir until dissolved.
3 Put glutinous rice flour into large bowl, add in water slowly to form smooth and elongated shape. Divide dough into 15 portions, and roll into round shape. Use forefinger to make a hole, wrap with some red bean paste, seal up the opening and roll into round shape.
4 Boil half pot of water. Add in rice balls, boil until floating and swell, and drain. Add rice balls into sweet soup, and serve.

核桃露花生湯圓
Peanut Rice Balls in Walnut Sweet Soup

數量：4 人份 / serves

烹製時間：60 分鐘 / mins

材料

湯料

- 去衣核桃 100 克
- 白米 40 克
- 清水 1000 ～ 1200 克
- 砂糖 100 克
- 牛奶 1/2 杯

湯圓料

- 糯米粉 150 克
- 清水 140 克（視情況加減）
- 花生餡 200 克

Ingredients

Walnut sweet soup

- 100g skinless walnut
- 40g white rice
- 1000 ～ 1200g water
- 100g sugar
- 1/2 cup milk

Rice balls

- 150g glutinous rice flour
- 140g water
- 200g peanut paste

做法

1. 花生餡分 20 等份，搓圓，放入冰箱定型，備用。
2. 核桃用白鍋烘香。
3. 白米用適量清水浸泡 3 小時，瀝乾水分。
4. 烘香核桃、白米加入部分清水用攪拌機打成核桃米漿。
5. 煮滾餘下的清水，加入核桃米漿煮滾，加入砂糖拌至糖溶解，加入牛奶拌勻。
6. 糯米粉放在大碗中，慢慢加入清水，搓揉成軟滑的長條，再分切成 20 等份，分別搓圓，用食指開窩，包花生餡，收口捏緊，搓圓。
7. 煮滾半鍋清水，放入湯圓煮至浮起脹身，撈起，將湯圓放入核桃露中，即可享用。

Methods

1. Divide peanut paste into 20 portions, roll to round shape, and store in refrigerator.
2. Roast walnut in wok to frangrant.
3. Soak white rice in water for 3 hrs, and drain.
4. Blend roasted walnut, rice and water in blender to form walnut rice mixture.
5. Boil the remaining water, add in walnut rice mixture, and boil. When boiled, add in sugar, and stir until dissolved. Add in milk, and stir well.
6. Put glutinous rice flour into large bowl, add in water slowly to form smooth and elongated shape. Divide dough into 20 portions, and roll into round shape. Use forefinger to make a hole, wrap with peanut paste, seal up the opening and roll into round shape.
7. Boil half pot of water. Add in rice balls, boil until floating and swell, and drain. Add rice balls into walnut sweet soup, and serve.

TIPS:

核桃可以放入烤箱用 120℃烤約 8 分鐘至香脆。

You may bake walnut in oven at 120℃ for about 8 mins to make it crispy.

黑白配

Black in White

數量：4 ～ 6 人份 / serves

烹製時間：60 分鐘 / mins

材料

湯料

黃豆 200 克
清水 900 克
砂糖 適量

湯圓料

・紫米湯圓

糯米粉 90 克
紫米水 75 ～ 80 克

Ingredients

Soymilk

200g soybean
900g water
Some sugar

Purple rice balls

90g glutinous rice flour
75 ～ 80g purple rice water

做法

1 黃豆泡水最少 4 小時，瀝乾。
2 黃豆與清水分幾次放入攪拌機中打成漿。然後用紗布過濾 3 次，擠出生豆漿。
3 生豆漿用中火煮滾，再轉小火煮約 30 分鐘，不要蓋鍋蓋，因為豆漿很容易溢出，要不時攪拌一下以免黏鍋底，煮滾後關火，調糖。
4 紫米水煮至和暖，加入糯米粉中搓揉成軟滑的粉糰，切粒，搓揉成小丸子。
5 燒滾半鍋清水，放入全部小丸子煮至熟，撈起，與豆漿一同拌吃。

Methods

1 Soak soybean for at least 4 hrs. Drain.
2 Pour soybean and water into blender by batches to form soymilk. Sieve the mixture with cheese cloth for 3 times and recover the soymilk.
3 Boil raw soymilk over medium heat, then turn to low heat and boil for about 30 mins. Do not cover when boiling as soymilk easily spills. Stir when boiling to avoid soymilk sticks to the bottom of pot. When boiled, turn heat off, and add some sugar.
4 Boil purple rice water until warm. Add in glutinous rice flour, mix to form smooth dough, and divide into small portions. Roll into round shape.
5 Boil half pot of water. Add in rice balls and boil until well done. Drain, and serve with soymilk.

TIPS:

泡豆時間因應天氣溫度而有差異，夏季 4 小時，春秋季 6 小時，冬季 8 小時，時間只供參考，最終都視乎實際環境而定。

在煮豆漿時會冒出很多泡沫，容易溢出，所以不要蓋鍋蓋及不時攪拌一下，以防黏鍋底。

The duration of soaking soybean varies in different temperatures. In summer, the soaking time is 4 hrs; in spring; the soaking time is 6 hrs; and in winter, the soaking time is 8 hrs. The time suggested is for reference only. It depends on the actual situation.

Lots of bubbles appear when boiling soymilk, and milk easily spills. Therefore, it is unnecessary to cover when boiling and please stir to avoid soymilk sticking to the bottom of pot.

三色湯圓

Tricolor Rice Balls

數量：*4* ～ *6* 人份 / serves

烹製時間：*40* 分鐘 / mins

材料

湯圓料

・巧克力湯圓

糯米粉 65 克
巧克力粉 1 茶匙
清水 55 克
鳳梨餡 45 克

・番薯湯圓

熱番薯 25 克 （去皮切片）
糯米粉 40 克
清水 55 克
綠豆餡 45 克

・原味湯圓

糯米粉 65 克
清水 55 克
芝麻餡 45 克

Ingredients

Chocolate rice balls

65g glutinous rice flour
1tsp cocoa powder
55g water
45g pineapple filling

Sweet potato rice balls

25g sweet potato (peel and slice)
40g glutinous rice flour
55g water
45g green bean paste

Original flavor rice balls

65g glutinous rice flour
55g water
45g black sesame paste

做法

1. 糯米粉放入大碗中，加入巧克力粉及清水搓揉成粉糰，再分成3等份，搓圓，用食指開窩，包入鳳梨餡，收口捏緊，搓圓。
2. 糯米粉放入大碗中，加熱番薯及清水搓揉成粉糰，再分成 3 等份，搓圓，用食指開窩，包入綠豆餡，收口捏緊，搓圓。
3. 糯米粉入大碗中，加清水搓揉成粉糰，分成 3 等份，搓圓，用食指開窩，包入芝麻餡，收口捏緊，搓圓。
4. 燒滾半鍋清水，放湯圓煮至熟，撈起。可隨意加糖水或沾椰絲進食。

Methods

1. Put glutinous rice flour into large bowl, add in cocoa powder and water to form smooth dough. Divide dough into 3 portions, and roll into round shape. Use forefinger to make a hole, wrap with pineapple filling, seal up the opening and roll into round shape.
2. Put glutinous rice flour into large bowl, add in hot sweet potato paste and water to form smooth dough. Divide dough into 3 portions, and roll into round shape. Use forefinger to make a hole, wrap with green bean paste, seal up the opening and roll into round shape.
3. Put glutinous rice flour into large bowl, add in water to form smooth dough. Divide dough into 3 portions, and roll into round shape. Use forefinger to make a hole, wrap with black sesame paste, seal up the opening and roll into round shape.
4. Boil half pot of water, add in rice balls and boil until well done. Drain and serve when hot. Rice balls can be served with sweet soup or dipping with shredded coconut.

紅豆草莓湯圓

Strawberry Rice Balls in Red Bean Sweet Soup

數量：4 ～ 6 人份 / serves

烹製時間：40 分鐘 / mins

材料

湯料（紅豆糖水）

紅豆 300 克
清水 900 克
冰糖 適量

湯圓料

・草莓湯圓

草莓果泥 90 克
糯米粉 150 克
清水 45 克（視情況加減）
草莓 4 ～ 5 顆

Ingredients

Sweet soup

300g red bean
900g water
some crystal sugar

Rice balls

90g strawberry puree
150g glutinous flour
45g water
4 ～ 5 strawberries

做法

1 紅豆沖洗後用清水浸泡 4 ～ 6 小時，濾掉水分，加清水煮滾，然後調中小火煮約 1 小時，熄火燜 15 分鐘，再開火煮至熟，加入冰糖拌至糖溶解。

2 草莓泡淡鹽水，沖洗乾淨，去蒂切粒，備用。

3 糯米粉加入草莓果泥拌勻，視情況加入適量清水搓揉成軟滑的長條，分切約 20 粒，分別搓圓，包入草莓粒，收口捏緊，搓圓。

4 燒滾半鍋清水，放入湯圓煮至浮起及脹身，撈出湯圓，盛入紅豆湯水中，即可享用。

Methods

1 Wash red bean and soak for 4 to 6 hrs. Drain. Cook until boiled, adjust the heat to medium low, simmer for 1 hr, off heat and keep for 15 mins. Re-cook until the beans become very soft. Add in crystal sugar and stir until sugar dissolved.

2 Soak strawberries in light salt water, then rinse. Cut off stem and dice. Set aside.

3 Add glutinous flour to strawberry puree, stir well. Add water if needed. Roll to form smooth and elongated dough. Divide into 20 portions, and roll into round shape. Use forefinger to make a hole, wrap with strawberry pieces, seal up the opening and knead into round shape.

4 Boil half pot of water. Add in rice balls, boil until floating and swell. Drain and put rice balls into red bean sweet soup. Serve.

TIPS:

浸泡紅豆要視天氣而定，夏天大約 3 ～ 4 小時，冬天 6 ～ 8 小時或泡過夜。
Time for soaking red beans depends on the temperature. It takes 3 ～ 4 hours in summer and 6 ～ 8 hours or overnight in winter.

薑汁南瓜湯圓
Pumpkin Rice Balls in Ginger Sweet Soup

數量：**4 ～ 6** 人份 / serves

烹製時間：**40** 分鐘 / mins

材料

南瓜 140 克 （去皮）
糯米粉 160 克
片糖 1 塊 （做餡）
片糖 100 克
薑汁 2 湯匙
清水 900 克

Ingredients

140g pumpkin (peeled)
160g glutinous rice flour
1pc lump sugar (as fillings)
100g lump sugar
2 tbsp ginger juice
900g water

做法

1 片糖 1 塊，用刀切粒，再挑選 20 粒大小適中的，備用。
2 將適量清水煮滾，加入薑汁、片糖煮成糖水，備用。
3 南瓜肉切片，大火蒸約 10 分鐘，趁熱將南瓜壓成泥狀，加入糯米粉拌勻，並搓揉成軟滑有光澤的長條。
4 再分切成20等份，每份包入一粒片糖，搓圓。
5 燒滾半鍋清水，放入南瓜湯圓煮至浮起及漲身，撈出湯圓，放入薑汁糖水內，即可享用。

Methods

1 Dice 1 pc of lump sugar. Select 20 pcs in proper size, and set aside.
2 Boil water, and add in ginger juice and lump sugar to make sweet soup. Set aside.
3 Slice pumpkin and steam over high heat for about 10 mins. When pumpkin is hot, mash it to make puree. Add in glutinous rice flour and mix well. Roll to form smooth and elongated dough.
4 Divide into 20 portions. Wrap in 1 pc lump sugar, and roll to round shape.
5 Boil half pot of water. Add in pumpkin rice balls, boil until floating and swell, and drain. Add into ginger juice sweet soup, and serve.

TIPS:

蒸起的南瓜含水量高，可以先加入一部分南瓜肉搓成泥，再視實際情況加減，太乾可以加南瓜泥，太濕可以加少許糯米粉。

Pumpkin contains large amount of water. You may add some pumpkin puree into glutinous rice flour and mix together first. Depending on the actual situation, add more pumpkin puree if too dry whereas add more glutinous rice flour if too wet.

喳咋南瓜湯圓

Pumpkin Rice Balls in Bubur Cha Cha

數量：4 ～ 6 人份 / serves
烹製時間：60 分鐘 / mins

材料

湯料

喳咋料 1 包
椰漿 400 克
芋頭 200 克
砂糖 100 克
清水 1000 克

湯圓料

・南瓜湯圓

南瓜 135 克（去皮）
糯米粉 145 克

Ingredients

1 pack bubur cha cha ingredients
400g coconut milk
200g taro
100g sugar
1000g water

Pumpkin rice balls

135g pumpkin (peeled)
145g glutinous rice flour

做法

1 芋頭去皮切粒；喳咋料先取出三角豆泡熱水 1 小時，連同其餘喳咋料沖水、瀝乾。

2 芋頭、喳咋料加清水一起煮滾，然後調中小火煮至熟，加入椰漿、砂糖拌至糖溶解。

3 南瓜肉切片，大火蒸約 10 分鐘，趁熱將南瓜壓成泥狀，加入糯米粉拌勻，並搓揉成軟滑有光澤的長條，再分切成 24 等份，搓圓。

4 燒滾半鍋清水，放入湯圓煮至浮起及脹身，撈出湯圓，放入喳咋內，趁熱享用。

Methods

1 Peel taro and dice. Get chickpea from cha cha dessert ingredients first and soak in hot water for 1 hr. Wash with other cha cha dessert ingredients, and drain.

2 Boil taro, cha cha dessert ingredients and water for a while. Turn to medium low heat and cook until sticky and well done. Add in coconut milk and sugar. Stir until sugar is dissolved.

3 Slice pumpkin and steam over high heat for about 10 mins. When pumpkin is hot, mash it to make puree. Add in glutinous rice flour and mix well. Roll to form smooth and elongated dough. Divide into 24 portions and roll to round shape.

4 Boil half pot of water. Add in rice balls, boil until floating and swell, and drain. Add into cha cha sweet soup, and serve when hot.

紫米芝麻糊湯圓

Rice Balls with Purple Rice and Sesame Sweet Soup

數量：4 ～ 6 人份 / serves
烹製時間：40 分鐘 / mins

材料

湯料

黑芝麻 150 克
白米 40 克
砂糖 100 克
清水 1500 克

湯圓料

・紅色小丸子

糯米粉 90 克
食用紅色素 少許
清水 80 克

Ingredients

150g black sesame
40g white rice
100g sugar
1500g water

Red rice balls

90g glutinous rice flour
1 drop red edible coloring
80g water

做法

1 白米用適量清水浸泡 3 小時，然後瀝乾水分。
2 白鍋烘香黑芝麻連同白米放入攪拌機內，加入約 400g 清水用打成芝麻漿。
3 然後用紗布過濾，再加餘下清水用中小火邊煮邊攪拌至煮滾，關火，調糖。
4 糯米粉加紅色素少許拌勻，搓揉成軟滑有光澤的長條，切成小丸子，搓圓。
5 燒滾半鍋清水，放入湯圓煮至浮起及脹身，撈出湯圓，放入芝麻糊內，趁熱享用。

Methods

1 Soak white rice in water for 3 hrs, and drain.
2 Roast black sesame in wok. Put it with white rice into blender. Add in 400g water and blend to form sesame milk.
3 Sieve by cheese cloth, add in the remaining water and boil over medium to low heat until well done. Remember to stir well during boiling. Turn heat off, and add sugar.
4 Add little red edible coloring into glutinous rice flour. Roll to form smooth and elongated dough, divide and roll to form round shape.
5 Boil half pot of water. Add in rice balls, boil until floating and swell, and drain. Add into sesame sweet soup, and serve when hot.

TIPS:

現在很多高速攪拌機可以不用濾芝麻渣，非常方便。
If high speed blender is used, you don't need to sieve the blended sesame milk.

紅豆綠泥湯圓

Green Bean Rice Balls in Red Bean Sweet Soup

數量：**4 ～ 6** 人份 / serves

烹製時間：**60** 分鐘 / mins

材料

湯料

- 紅豆 300 克
- 清水 900 克
- 冰糖 適量

湯圓料

- 糯米粉 150 克
- 溫水 130 克
- 綠豆餡 200 克

Ingredients

Red bean sweet soup

300g red bean
900g water
Some crystal sugar

Rice balls

150g glutinous rice flour
130g warm water
200g green bean paste

做法

1. 紅豆沖洗後用清水浸泡 3 小時，濾掉水分，加清水煮滾，然後調中小火煮約 1 小時，熄火焖 15 分鐘，再開火煮至熟，加入冰糖拌至糖溶解。
2. 綠豆餡分為 20 等份，搓圓，蓋上保鮮膜，再放入冰箱冷凍定型。
3. 糯米粉加入溫水拌勻，搓揉成軟滑的長條，分切成 20 等份，分別搓圓，用食指開窩，包入綠豆餡，收口捏緊，搓圓。
4. 燒滾半鍋清水，放入湯圓煮至浮起及脹身，撈出湯圓，盛入紅豆湯水中。

Methods

1. Wash red bean and soak for 3 hrs. Drain, add in water and boil. Turn to medium heat and boil for about 1 hr. Turn heat off and braise for 15 mins. Turn heat on and boil until sticky and well done. Add in crystal sugar and stir until sugar is dissolved.
2. Divide green bean paste into 20 portions. Roll round, cover with plastic wrap and store in refrigerator.
3. Put glutinous rice flour into large bowl, add in warm water and mix well. Roll to form smooth and elongated dough. Divide dough into 20 portions, and roll into round shape. Use forefinger to make a hole, wrap in green bean paste, seal up the opening and roll into round shape.
4. Boil half pot of water. Add in rice balls, boil until floating and swell, and drain. Add into red bean sweet soup, and serve when hot.

TIPS:

紅豆糖水可以加入陳皮一起煮，陳皮一角，用清水浸軟，刮掉果皮內的纖維連同紅豆一起入鍋內煮，煮滾後取出切細絲再放回糖水內。

You may boil red bean sweet soup with dried citrus peel. Soak dried citrus peel in water to make it soft. Rub and remove fibre of dried citrus peel, and boil with red bean in pot. After boiled, take dried citrus peel out and shred, then put back into sweet soup.

杏香紫雲湯圓

Green Bean Rice Balls in Red Bean Sweet Soup

數量：*4 ～ 6* 人份 / serves

烹製時間：*40* 分鐘 / mins

TIPS:

紫米一小撮，加清水泡一會，濾出紫米水便是天然顏色染劑。

Soak some purple rice in water for a while, and drain. The soaked purple rice water can be used as natural coloring.

材料

湯料（杏仁茶）
- 南杏 200 克
- 北杏 10 克
- 白米 60 克
- 清水 1000 克
- 冰糖 90 克

湯圓料

・深紫湯圓
- 糯米粉 95 克
- 紫米水 85 克

・淺紫湯圓
- 芋泥 50 克
- 糯米粉 45 克
- 清水 85 克

・原味芋泥湯圓
- 糯米粉 95 克
- 溫水 85 克
- 芋泥餡 120 克

做法

1. 南、北杏、白米用適量清水浸泡最少 4 小時，然後瀝乾水分。
2. 將糖水材料內的清水與南北杏及白米分數次放入攪拌機內打磨成杏仁米漿，再用紗布濾渣留汁。
3. 杏仁汁用中小火煮滾，加入冰糖拌至糖溶解，而在煮的過程中要不時攪拌一下以免黏鍋。
4. 淺紫：芋泥蒸熱加入糯米粉、清水，搓揉成軟滑的粉糰，備用。
5. 深紫：紫米水煮至和暖，加入糯米粉中搓揉成軟滑的粉糰，備用。
6. 原味：芋泥餡分成 12 份；糯米粉加入溫水搓揉成軟滑的長條粉糰，分切 12 份，包餡，搓圓。
7. 淺紫及深紫粉糰分別搓成長條，然後將兩條扭在一起，分切成小丸子，搓圓。
8. 燒滾半鍋清水，先煮小丸子，煮熟後撈起，放入湯圓煮至熟，將煮熟的大小湯圓放入碗內，再加入杏仁茶，即可享用。

Ingredients

Light purple rice ball
- 50g taro paste
- 45g glutinous rice flour
- 85g water

Taro rice balls
- 95g glutinous rice flour
- 85g warm water
- 120g taro filling

Dark purple rice balls
- 95g glutinous rice flour
- 85g purple rice water

Almond sweet soup
- 200g apricot kernel
- 10g bitter almond
- 60g white rice
- 1000g water
- 90g crystal sugar

Methods

1. Soak apricot kernel, bitter almond and white rice for at least 4 hrs, and drain.
2. Pour part of water, apricot kernel, bitter almond and white rice into blender to make almond rice milk. Sieve by cheese cloth and recall the rice milk.
3. Cook almond rice milk over medium heat. Add in crystal sugar and boil until sugar is dissolved. Stir frequently during boiling to avoid the mixture sticking to the pot.
4. Light purple dough: Steam taro until hot, add in glutinous rice flour and water. Roll to form smooth dough, and set aside.
5. Dark purple dough: Boil purple rice water until warm, add in glutinous rice flour, roll to form smooth dough, and set aside.
6. Taro rice ball: Divide taro filling into 12 portions. Mix glutinous rice flour and warm water, and roll to form smooth and elongated dough. Divide into 12 portions, wrap in fillings and roll to round shape.
7. Roll light and dark purple dough into elongated shape separately. Then twist them together, divide and roll to round shape.
8. Boil half pot of water. Add in small rice balls first, cook until well done, and drain. Add in large rice balls and boil until well done. Pour small and large rice balls into bowl, add in almond tea and serve.

喳咋湯圓

Rice balls with Bubur Cha Cha Dessert

數量：4 ～ 6 人份 / serves
烹製時間：60 分鐘 / mins

材料

湯料（糖水）

喳咋料 1 包
椰漿 400 克
芋頭 200 克
砂糖 100 克
清水 適量

湯圓料

・南瓜湯圓

南瓜 135 克 （去皮）
糯米粉 145 克

Ingredients

Sweet soup

1 pack of cha cha dessert ingredients
400g coconut milk
200g taro
100g sugar
appropriate amount of water

Pumpkin rice balls

135g pumpkin (peeled)
145g glutinous rice flour

做法

1. 芋頭去皮切粒；喳咋料先取出三角豆泡熱水 1 小時，連同其餘喳咋料沖水、瀝乾。
2. 芋頭、喳咋料加清水一起煮滾，然後調中小火煮至熟，加入椰漿、砂糖拌至糖溶解。
3. 南瓜肉切片，大火蒸約 10 分鐘，趁熱將南瓜壓成泥狀，加入糯米粉拌勻，並搓揉成軟滑有光澤的長條，再分切成 20 等份，搓圓。
4. 燒滾半鍋清水，放入湯圓煮至浮起及脹身，撈出湯圓，放入喳咋內，趁熱享用。

Methods

1. Peel taro and dice. Get chickpea from cha cha dessert ingredients first and soak in hot water for 1 hr. Wash with other cha cha dessert ingredients, and drain.
2. Boil taro, cha cha dessert ingredients and water for a while. Turn to medium low heat and boil until sticky and well done. Add in coconut milk and sugar. Stir until sugar is dissolved.
3. Slice pumpkin and steam over high heat for about 10 mins. When pumpkin is hot, mash it to make puree. Add in glutinous rice flour and mix well. Roll to form smooth and elongated dough. Divide into 20 portions and roll to round shape.
4. Boil half pot of water. Add in rice balls, boil until floating and swell, and drain. Add into cha cha sweet soup, and serve when hot.

PART 3
家鄉風味的鹹湯圓

鮮肉湯圓

Pork Rice Balls in Mushroom Soup

數量：3 ～ 5 人份 / serves

烹製時間：30 分鐘 / mins

TIPS:

湯圓不宜包得太多餡，否則煮至湯圓皮過熟而肉餡有可能未煮熟。

Remember not to wrap too much fillings. Otherwise, rice ball dough will be overcooked whereas fillings will be undercooked.

材料

湯料

- 高湯 1000 克
- 冬菇 5 朵
- 菜心 3 棵
- 鹽和胡椒粉適量

肉醃料

- 鹽 1/4 茶匙
- 胡椒粉少許
- 太白粉 1 茶匙

餡料

- 絞肉 200 克
- 蝦米 1 小撮
- 紅葱頭 3～4 粒

湯圓料

- 糯米粉 180 克
- 熱水 40 克
- 清水 100 克

Ingredients

Marinades

- 1/4 tsp salt
- Some pepper
- 1 tsp cornstarch

Soup

- 1000g broth
- 5 pcs Chinese mushroom
- 3 choysum
- Salt & pepper

Filling

- 200g minced pork
- Some dried shrimps
- 3～4 shallots

Rice balls

- 180g glutinous rice flour
- 40g hot water
- 100g water

做法

1. 冬菇泡軟，去蒂。蝦米泡軟，略為切碎；紅葱頭剁泥，備用。絞肉加入醃料拌勻，醃 15 分鐘。
2. 燒熱鍋，下多一點油，加入紅葱頭和蝦米爆香，盛起葱泥和蝦米後，留油在鍋內。
3. 熱鍋下醃好的豬絞肉，炒至肉色變白，把葱泥和蝦米回鍋，炒勻，試味並按需要加鹽調味，以少許太白粉水勾芡，拌勻成餡料，待涼後放入冰箱 1～2 小時。
4. 高湯加入冬菇煮滾，下菜心，以鹽和胡椒粉調味，備用。
5. 糯米粉放入大碗中，加入熱水拌勻，再加入清水搓揉成粉糰。
6. 粉糰分成 15 等份，分別搓圓，用大拇指開窩，包入適量餡料，收口捏緊，搓圓。
7. 燒滾半鍋清水，放入湯圓煮滾，調小火煮至湯圓浮起及漲身。
8. 將煮好的湯圓放入冬菇湯中，即可食用。

Methods

1. Soak Chinese mushroom to tender, then cut into slices. Soak dried shrimps, drain and roughly chop. Fine chop shallots. Marinate pork with marinades for 15 minutes. Set aside.
2. Heat wok with oil, sauté chopped dried shrimp and shallot until fragrant. Remove the ingredients and keep the remained oil.
3. Add minced pork to the hot oil and stir-fry until the pork changes colour. Add dried shrimp and shallot and stir well. Taste and add salt if needed, thicken with cornstarch solution. When cool, put in refrigerator for half day before use.
4. Add Chinese mushroom into broth and boil. Then add in choysum, salt and pepper.
5. Put glutinous rice flour into large bowl, add in hot water and mix well. Then add in water, and mix well to make dough.
6. Divide dough into 15 portions, and roll into round shape. Use forefinger to make a hole, wrap with some pork fillings, seal up the opening and roll into round shape.
7. Boil half pot of water, add in rice balls, and cook until floating. Add in 1/2 bowl of water and boil until rice balls swell.
8. Put cooked rice balls into broth, and serve.

鹹雞湯圓
Salty Chicken Rice Balls

數量：4 ～ 6 人份 / serves
烹製時間：60 分鐘 / mins

TIPS:

蒸雞時留下的雞汁可以拿來炒菜或是將浮面的油脂隔掉，放入高湯內一起煮。
The chicken sauce left after steaming of chicken can be used for stir frying of vegetables. Or remove the fat on the surface of chicken sauce, then put into broth and cook together.

材料

湯料

- 高湯 1000 克
- 香菇 4 朵（切片）
- 蝦米 1 湯匙（泡軟）
- 蒜頭 2 ～ 3 瓣
- 葱花 適量
- 鹽 適量
- 胡椒粉 適量

湯圓料

- 糯米粉 165 克
- 清水 150 克

鹹雞

- 冰鮮雞 1 隻（大約 960 克）
- 薑 3 ～ 4 片
- 葱 2 ～ 3 條
- 鹽 2 湯匙

Ingredients

Rice balls

- 165g glutinous rice flour
- 150g water

Soup

- 1000g broth
- 4 pcs Chinese mushroom (sliced)
- 1 tbsp dried shrimp (soaked)
- 2 ～ 3 pcs garlic
- Some chopped spring onion
- Some salt
- Some pepper

Salty chicken

- 1 chicken (about 960g)
- 3 ～ 4 pcs ginger
- 2 ～ 3 pcs spring onion
- 2 tbsp salt

做法

1. 香菇泡軟，切片；蝦米泡軟；蒜頭剁泥，備用。
2. 冰鮮雞清洗乾淨，瀝乾，用廚房紙將雞隻內外的水分抹乾。用鹽抹遍雞身內外，醃約 2 ～ 3 小時。將薑葱塞入雞腔內，大火蒸約 25 分鐘至熟，取出，攤涼，切塊。
3. 高湯放入鍋內煮滾，加入鹽、胡椒粉調味，備用。
4. 起鍋下油，放入蒜泥、蝦米爆香，再加入香菇片炒勻。
5. 將蝦米香菇配料放入高湯內，備用。
6. 糯米粉放入大碗中，加入清水拌勻，再搓揉成粉糰。將粉糰搓長，分成 30 等份的小丸子，分別再搓圓。
7. 煮滾半鍋清水，放入小丸子煮至浮起及脹身。
8. 將煮好的湯圓放入高湯內，撒上葱花，即可食用。

Methods

1. Soak Chinese mushrooms to soft, and slice. Soak dried shrimp to soft. Fine chop garlic. Set aside.
2. Wash chicken and drain. Pat dry with kitchen paper. Rub the chicken skin and inside with salt and marinate for 2 ～ 3 hours. Put ginger and green onion into chicken, steam over high heat for about 25 minutes until well done. Remove, let cool and chop.
3. Boil broth, add in salt and pepper. Set aside.
4. Heat wok with oil. Sauté garlic puree and dried shrimp. Then add in sliced Chinese mushroom, and stir fry.
5. Put dried shrimp and mushroom into broth. Set aside.
6. Put glutinous rice flour into large bowl, add in water and mix well. Then roll into dough and elongated shape. Divide into 30 portions and roll into round shape.
7. Boil half pot of water, add in rice balls, boil until floating and swell.
8. Put cooked rice balls into broth. Sprinkle chopped green onion and serve.

雪菜火鴨絲小丸子

Rice Balls with Preserved Vegetables and Shredded Roasted Duck

數量：4 ～ 6 人份 / serves
烹製時間：45 分鐘 / mins

材料

湯料

- 燒鴨 1/4 隻
- 雪菜（雪裡蕻） 1 棵
- 薑 3 片
- 高湯 800 克
- 鹽 少許
- 胡椒粉 少許

湯圓料

- 糯米粉 180 克
- 溫水 140 克（視情況加減）

Ingredients

- 1/4 pc roasted duck
- 1 stalk preserved vegetables
- 3 pcs ginger
- 800g broth
- Some salt
- Some pepper

Rice balls

- 180g glutinous rice flour
- 140 warm water
（adding amount is subject to actual situation）

做法

1. 雪菜用淡鹽水浸泡 30 分鐘，沖洗乾淨，瀝乾，切掉根部然後切丁。
2. 燒鴨起骨切鴨絲；將鴨骨加入高湯內煮滾，下鹽、胡椒粉調味。
3. 燒油鍋，下油爆香薑片，加入鴨絲炒片刻，加雪菜炒勻，盛起，備用。
4. 糯米粉放入大碗中，加溫水拌勻，搓揉成粉糰，再分切成小糰子，搓圓。
5. 煮滾半鍋清水，放入小丸子煮至浮起，撈起小丸子，與雪菜火鴨絲一同放入高湯內煮片刻，即可享用。

Methods

1. Soak preserved vegetables in light salt water for 30 minutes. Wash and drain. Cut away the root and dice.
2. Remove the bone of duck, and shred the meat. Put duck bone into broth and boil. Add in salt and pepper.
3. Heat wok with oil and Sauté sliced ginger. Add in shredded duck meat, and stir fry for a while. Then add in preserved vegetables, stir fry and set aside.
4. Put glutinous rice flour into large bowl, add in warm water and mix well. Then roll into dough, divide into small portions and roll into round shape.
5. Boil half pot of water, add in rice balls, boil until floating and swell. Drain rice balls, put it together with preserved vegetables and shredded roasted duck into broth, boil for a while and serve.

味噌豆腐湯圓
Miso Tofu Rice Balls

數量：3～4人份 / serves
烹製時間：25分鐘 / mins

材料

湯料

- 鰹魚湯粉 1 包（10g）
- 清水 800 克
- 嫩豆腐 1/2 塊
- 海藻 6 克
- 白味噌 2～3 湯匙
- 溫水 1 湯匙
- 葱花 適量

湯圓料

・豆腐湯圓

- 嫩豆腐 半塊
- 糯米粉 180 克
- 溫水 適量

Ingredients

Soup

1 pack striped tuna soup powder (10g)
800g water
1/2 pc soft tofu
6g seaweed
2～3 tbsp white miso
1 tbsp warm water
Some chopped spring onion

Tofu rice balls

1/2 pc soft tofu
180g glutinous rice flour
Some warm water

做法

1. 糯米粉與嫩豆腐放入大碗中，用手搓揉至兩者混合，如果太硬加溫水，太軟加糯米粉，搓揉成粉糰，再分切成小丸子，搓圓。
2. 煮滾半鍋清水，放入小丸子煮至浮起及脹身，撈起小丸子，浸冰水。
3. 海藻浸清水 15 分鐘後洗淨；豆腐切塊。
4. 味噌用溫水先行調開。
5. 清水煮滾，放入鰹魚湯粉煮至溶解；加入味噌水拌勻煮滾，試味，加入海藻、豆腐、湯圓煮滾，撒下葱花，即可食用。

Methods

1. Mix glutinous rice flour and soft tofu in large bowl by hand. If it is too hard, you may add some warm water. If too soft, you may add some glutinous rice flour. Roll to form dough. Divide and roll to form round shape.
2. Boil half pot of water. Add in rice balls and boil until floating and swell. Drain, and put into iced water.
3. Soak seaweed in water for 15 mins and wash. Dice tofu.
4. Dilute miso with warm water.
5. Boil water, add in striped tuna soup powder and cook until dissolved. Add in miso water, stir, boil and taste. Add in seaweed, tofu and rice balls, and boil. Sprinkle chopped spring onion and serve.

TIPS:

味噌即麵豉湯，有白、紅及咖啡色之分，味道各有不同，以白味噌的口味較淡一些，選擇時要小心留意。

Different kinds of miso (i.e. white, red and brown miso) have different tastes / flavors (e.g. white miso gives light flavor). Therefore, be cautions when choosing miso.

香芹鹹湯圓

Salty Maize Rice Balls

數量：3～5人份 / serves
烹製時間：25 分鐘 / mins

材料

湯料

香菇 4 朵
粉絲 1 小紮
高湯 1000 克
芹菜 適量
鹽 適量
胡椒粉 適量

湯圓料

玉米粒 30 克
糯米粉 180 克
熱水 40 克
清水 100 克

Ingredients

Soup

4 pcs Chinese mushroom
1 stalk mengbean vermicelli
1000g broth
Some Chinese celery
Some salt
Some pepper

Rice balls

30g corn
180g glutinous rice flour
40g hot water
100g water

做法

1. 玉米粒用廚房紙吸乾水分；粉絲泡軟，切段；芹菜切段，備用。
2. 糯米粉放入大碗中，加入熱水拌勻，再加入清水搓揉成粉糰，分切 15 等份。
3. 分別搓圓，用食指開窩，包入適量玉米粒，收口捏緊，搓圓。
4. 燒滾半鍋清水，放入湯圓煮至浮起，再加入半碗清水煮至湯圓脹身。
5. 高湯加入香菇煮滾，下粉絲、芹菜，以鹽和胡椒粉調味。
6. 將煮好的湯圓放入高湯內，再煮滾，即可食用。

Methods

1. Use kitchen paper to drain maize. Soak mengbean vermicelli to make it soft, and cut. Cut Chinese Celery in short, and set aside.
2. Put glutinous rice flour into large bowl, add in hot water, and mix well. Then add in water, roll to make dough, and divide into 15 portions.
3. Roll into round shape. Use forefinger to make a hole, wrap with some corns, seal up the opening and roll into round shape.
4. Boil half pot of water, add in rice balls, and cook until floating. Add in 1/2 bowl of water and boil until rice balls swell.
5. Add Chinese mushroom into broth and bring to a boil. Then add in mengbean vermicelli, celery, salt and pepper.
6. Put cooked rice balls into soup, bring to a boil. Serve in bowls.

魚肉碗仔翅湯圓

Rice Balls with Mashed Fish and Vermicelli Soup

數量： 3 ～ 5 人份 / serves

烹製時間：30 分鐘 / mins

材料

湯料

- 鯪魚肉 300 克
- 木耳絲 10 克
- 香菇 4 朵
- 粉絲 1 小紮
- 高湯 1000 克
- 雞蛋 1 顆
- 鹽 適量
- 胡椒粉 適量
- 麻油 少許
- 太白粉芡（太白粉 2 湯匙加清水 1 湯匙）

湯圓料

・小丸子

- 糯米粉 105 克
- 溫水 95 克

湯調味料

- 鹽 1/4 茶匙
- 胡椒粉 適量
- 太白粉 1 茶匙
- 沙拉油 1 茶匙（後下）

做法

1. 香菇泡軟，切片；木耳絲、粉絲分別浸軟。
2. 鯪魚肉加入醃料攪至有黏性。
3. 高湯煮滾，加入木耳、香菇煮滾，金屬湯匙沾水，將鯪魚肉逐次刮至高湯內，煮至再滾起，加入粉絲，下鹽、胡椒粉調味拌勻。
4. 將太白粉芡拌勻，左手下右手推（逐次加入再看情況，不要全部加入），拌勻煮滾後關火。
5. 在湯面加入麻油及打散的蛋液，再用湯匙推成蛋花。
6. 糯米粉放入大碗中，加入溫水拌勻，搓揉成粉糰，再分切成小丸子，搓圓。
7. 燒滾半鍋清水，放入小丸子煮至浮起及脹身，撈起小丸子加入魚翅湯內，趁熱享用。

Ingredients

Soup

- 300g dace fish meat
- 10g shredded black fungus
- 4 pcs Chinese mushroom
- 1 stalk mengbean vermicelli
- 1000g broth
- 1 pc egg
- Some salt
- Some pepper
- Some sesame oil
- thickening (mix 2 tbsp starch with 1 tbsp water)

Soup seasonings

- 1/4 tsp salt
- Some pepper
- 1 tsp corn starch
- 1 tsp oil (add last)

Rice balls

- 105g glutinous rice flour
- 95g warm water

Methods

1. Soak Chinese mushroom to make it soft, and shred. Soak black fungus and mengbean vermicelli separately.
2. Add marinades / seasonings into dace fish meat, and stir until sticky.
3. Boil broth, add in black fungus and Chinese mushroom. Dip metal spoon with water, use it to put little bit of dace fish meat into broth, and boil until floating. Add in Mengbean vemicelli, salt and pepper.
4. Stir Cornstarch, add thickening into broth gradually (do not add all). Switch off heat after mixing and well done.
5. Add in sesame oil and whisked egg. Use ladle to make egg drop / flower.
6. Put glutinous rice flour into large bowl, add in warm water, mix well, and roll to form dough. Divide into small portions, and roll to form round shape.
7. Boil half pot of water, add in rice balls, and cook until floating and swell. Drain rice balls, and put into mashed fish and vermicelli soup. Serve when it is hot.

生菜魚肉花生湯圓

Peanut Rice Balls with Mashed Fish and Lettuce

數量： **3 ～ 5** 人份 / serves

烹製時間：**40** 分鐘 / mins

材料

湯料

- 鯪魚肉 150 克
- 蝦皮 1 把
- 蘿蔔 1/2 棵
- 生菜絲 1 棵
- 高湯 適量
- 鹽 適量
- 胡椒粉 少許
- 麻油 少許

醃料

- 鹽 1/8 茶匙
- 胡椒粉 適量
- 玉米粉 1/2 茶匙
- 植物油 1/2 茶匙（後下）

湯圓料

・花生湯圓

- 糯米粉 100 克（視情況加減）
- 溫水 85 克
- 熟花生 20 ～ 30 粒

Ingredients

150g dace fish meat
Some dried tiny shrimp
1/2 pc carrot
1 pc lettuce (shredded)
Some broth
Some salt
Some pepper
Some sesame oil

Peanut rice balls

100g glutinous rice flour
85g warm water
20 ～ 30 pcs cooked peanut

Marinades

1/8 tsp salt
Some pepper
1/2 tsp Cornstarch
1/2 tsp oil (add last)

做法

1. 蝦皮沖洗一下；蘿蔔刨皮切條，放入滾水中汆燙；鯪魚肉加入醃料攪至有黏性。
2. 高湯加入蝦皮煮滾，金屬湯匙沾水，將鯪魚肉逐次刮至高湯內，加入蘿蔔條煮至滾起，再放入生菜絲，下鹽、胡椒粉，麻油調味。
3. 糯米粉放入大碗中，加入溫水拌勻，搓揉成粉糰，再分成 15 等份，每份包入 2 ～ 3 粒熟花生，收口及搓圓。
4. 燒滾半鍋清水，放入湯圓煮至浮起及脹身，撈起加入魚肉湯內，趁熱享用。

Methods

1. Wash dried tiny shrimp. Peel carrot, shred, and boil in hot water for a while. Add marinades into dace fish meat, and stir until sticky.
2. Put shrimp bran into broth and boil. Dip metal spoon with water, use it to put little bit of dace fish meat into broth, add in shredded carrot, and boil until floating. Then add in shredded lettuce, salt, pepper and sesame oil.
3. Put glutinous rice flour into large bowl, add in warm water, and mix well. Roll to form dough, divide into 15 portions, and wrap in 2 ～ 3 pcs cooked peanuts, seal up the opening and roll into round shape.
4. Boil half pot of water, add in rice balls, boil until floating and swell. Drain and put into fish soup. Serve when hot.

蘿蔔魚條小丸子

Rice Balls in Radish and Fish Finger Rice Balls

數量： 4 ～ 6 人份 / serves

烹製時間：40 分鐘 / mins

材料

湯料

鯪魚肉 150 克
蘿蔔 200 克（切塊）
娃娃菜 3 棵
香菇 3 朵（切絲）
魚湯 800 克
鹽 適量
胡椒粉 適量
麻油 少許

湯圓料

・黃色小丸子

南瓜泥 80 克
糯米粉 90 克

・紅色小丸子

糯米粉 105 克
食用紅色素 少許
清水 95 克

醃料

鹽 1/8 茶匙
胡椒粉 適量
太白粉 1/2 茶匙
植物油 1/2 茶匙（後下）

Ingredients

Soup

150g dace fish meat (minced)
200g radish (chopped)
3 baby cabbages
3 Chinese mushrooms (shredded)
800g fish soup
Some salt and pepper
Some sesame oil

Red rice balls

105g glutinous rice flour
1 drop red edible coloring
95g water

Yellow rice balls

80g pumpkin puree
90g glutinous rice flour

Marinades

1/8 tsp salt
Some pepper
1/2 tsp cornstarch
1/2 tsp oil (add last)

做法

1. 香菇泡軟切片；蘿蔔刨皮切條，放入滾水中汆燙；娃娃菜洗淨後切半，瀝乾。
2. 鯪魚肉加入醃料攪至有黏性，雙手沾水把魚肉弄成魚餅，熱鍋下油煎熟魚餅，切條，備用。加入魚湯內。
3. 南瓜泥蒸熱加糯米粉拌勻，搓揉成軟滑有光澤的長條，分切成小丸子，搓圓。
4. 糯米粉入大碗中，加色素少許並搓揉成軟滑有光澤的長條，切成小丸子搓圓。
5. 魚湯煮滾，加入香菇煮片刻，放入蘿蔔、娃娃菜、鯪魚條煮滾，下鹽、胡椒粉調味，備用。
6. 燒滾半鍋清水，放入小丸子煮至浮起及脹身，撈起小丸子加入魚湯內，加入麻油趁熱享用。

Methods

1. Soak Chinese mushroom and slice. Peel radish, cut into pieces, and blanch in boiling water. Wash baby cabbages and cut in half, drain to dry.
2. Stir minced dace fish meat with marinades until sticky. Wet both hands with water, make fish paste into cake shape. Heat wok with oil and pan fry fish cake to done. Slice and put into fish soup.
3. Steam pumpkin puree, then add in glutinous rice flour and mix well. Roll to form smooth and elongated shape. Dice and roll each to form round shape.
4. Put glutinous rice flour into large bowl. Add in a drop of red edible coloring. Mix to form smooth and elongated shape. Divide and roll to form rice balls.
5. Boil fish soup, add in Chinese mushroom and radish, until re-boiled, put fish slice in and add salt to taste.
6. Boil half pot of water, add in rice balls, boil until floating and swell. Drain and put into fish soup. Serve when hot.

魚湯寶圓

Rice Balls in Fish Soup

數量： 2 ～ 4 人份 / serves

烹製時間：45 分鐘 / mins

材料

湯料

- 小魚 6 條
- 大眼雞魚 1 條
- 粉絲 1 小紮
- 葱 1 棵（切段）
- 鹽 2 茶匙
- 薑片 3 片
- 清水 適量

湯圓料

- 蝦皮 1 把
- 絞肉 150 克
- 糯米粉 180 克
- 熱水 40 克
- 清水 100 克

醃料

- 鹽 1/4 茶匙
- 胡椒粉 少許
- 太白粉 1/2 茶匙

Ingredients

Rice balls

- 1 pc dried tiny shrimp
- 150g minced pork
- 180g glutinous rice flour
- 40g hot water
- 100g water

Marinades

- 1/4 tsp salt
- Some pepper
- 1/2 tsp cornstarch

Soup

- 6 pcs small fish
- 1 pc priacanthus hamrur
- red big eye
- 1 stalk mengbean vemicelli
- 1 stalk spring onion (shredded)
- 2 tsp salt
- 3 pcs ginger (sliced)
- Some water

做法

1. 蝦皮沖洗乾淨，瀝乾；絞肉加醃料拌至有黏性，放入冰箱半天才使用。
2. 小魚及大眼雞魚去內臟洗淨，用廚房紙吸乾水分，再用鹽將魚身內外抹勻，醃 15 分鐘。
3. 起鍋下油，爆香薑片，放入大眼雞魚煎至兩面金黃，取出，瀝油。
4. 小魚下鍋煎香，加入適量清水大火煮滾，滾起再加清水，分段加水，魚湯才會呈奶白色。
5. 取出小魚棄掉，魚湯加入粉絲，大眼雞魚煮滾，以鹽水調味，加入葱段。
6. 糯米粉放入大碗中，加入熱水拌勻，再加入清水搓揉成粉糰團，再分成 10 等份，每份包入絞肉餡料。
7. 燒滾半鍋清水，放入肉圓煮至浮起，再加入半碗清水煮至脹身，撈起湯圓加入魚湯內，趁熱享用。

Methods

1. Wash dried tiny shrimp and drain. Add marinades into minced pork and stir until sticky, then store in refrigerator for half day before use.
2. Wash and clean small fish and red big eye. Pat dry with kitchen paper. Rub fish (outer and inner) with salt, and marinate for 15 mins.
3. Heat wok with oil, and sauté with ginger. Pan fry red big eye until both sides are golden brown. Remove and drain.
4. Pan fry small fish. Add in water and boil over high heat. When boiled, add some water, and repeat this step to make fish soup milky.
5. Remove small fishes, add red big eye and mengbean vermicelli into the soup, boil for a while, add salt to taste. Put in spring onion.
6. Put glutinous rice flour into large bowl. Add in hot water and then cold water, Mix well to form a smooth dough. Roll to elongated shape. Divide into 10 portions, and wrap in pork filling.
7. Boil half pot of water, add in rice balls, boil until floating, add half bowl of cold water, cook until the rice balls swell. Drain and put into fish soup. Serve when hot.

三鮮鹹湯圓

Salty Rice Balls in Tasty Soup

數量： *6* 人份 / serves

烹製時間：*40* 分鐘 / mins

材料

湯料

- 香菇 6 朵
- 金菇 1 小把
- 銀耳（白木耳） 10 克
- 蘿蔔 200 克
- 豬肉 100 克
- 金華火腿 4 ～ 5 片
- 胡椒粉 少許
- 清雞湯 1000 克

湯圓料

- 糯米粉 120 克
- 清水 110 克

醃料

- 鹽 少許
- 胡椒粉 少許

Ingredients

Soup

6 pc Chinese mushroom
Bunch of Enoki mushroom
10g black fungus
200g carrot
100g pork
4 ～ 5 pcs Jinhua ham
1000g chicken soup

Rice Balls

120g glutinous rice flour
110g water

Marinades

Some salt
Some pepper

做法

1. 香菇泡軟，去蒂；銀耳泡軟；蘿蔔去皮切條，出水。
2. 金菇要在使用前才能沖洗及切去底部位置。
3. 金華火腿略為蒸熟；豬肉切片，用醃料醃 15 分鐘。
4. 湯圓料拌勻，搓揉成軟滑的長條，切粒，搓圓。
5. 燒滾半鍋清水，放入湯圓煮至浮起及脹身，撈出湯圓，分裝小碗。
6. 清雞湯加入豬肉煮滾，再加入其餘材料，以鹽調味。舀入盛有湯圓的碗中，即可享用。

Methods

1. Soak Chinese mushroom and remove stem. Soak black fungus to make soft. Peel carrot, shred and blanch.
2. Clean Enoki mushroom and cut away bottom part just before use.
3. Steam Jinhua ham little bit. Slice pork and marinate for 15 mins.
4. Mix rice balls ingredients well. Roll to form smooth and elongated shape. Divide and roll to round shape.
5. Boil half pot of water. Add in rice balls, boil until floating and swell, and drain. Place into small bowls.
6. Put pork into chicken soup. When boild, add in all other ingredients. Add little salt to taste. Scope into bowls with rice balls . Serve.

臘味魚肉鹹湯圓

Rice Balls with Chinese Preserved Meat and Dace Fish Balls

數量： 6 ～ 8 人份 / serves

烹製時間：60 分鐘 / mins

材料

湯料

- 臘腸 1 條
- 鯪魚肉 150 克
- 金華火腿 3 ～ 4 片
- 芫荽（香菜） 1 棵
- 清雞湯 1000 克
- 鹽 少許
- 胡椒粉 少許

湯圓料

- 糯米粉 120 克
- 清水 110 克

醃料

- 鹽 1/8 茶匙
- 胡椒粉 適量
- 太白粉 1/2 茶匙
- 植物油 1 茶匙 （後下）

Ingredients

Soup

1 pc Chinese preserved sausage
150g dace fish meat
3 ～ 4 pcs Jinhua ham
1 stalk coriander
1000g chicken soup
Some salt
Some pepper

Rice balls

120g glutinous rice flour
110g water

Marinades

1/8 tsp salt
Some pepper
1 tsp Cornstarch
1 tsp oil (add last)

做法

1. 臘腸蒸熟、切片。金華火腿略為蒸熟。
2. 芫荽洗淨後切小段。
3. 鯪魚肉加入醃料攪至有黏性，然後加油攪勻。
4. 湯圓料拌勻，搓揉成軟滑的長條，切粒，搓圓。
5. 燒滾半鍋清水，放入湯圓煮至浮起及脹身，撈出湯圓，備用。
6. 煮滾雞湯，金屬湯匙沾水將鯪魚肉逐次刮至雞湯內，煮至再滾起。下臘腸片、金華火腿片、芫荽及湯圓，用鹽調味後即可食用。

Methods

1. Steam Chinese preserved sausage until well done, and slice. Steam Jinhua ham for a while.
2. Wash coriander and chop.
3. Add marinades into dace fish meat until sticky, then stir in oil.
4. Mix rice balls ingredients well. Roll to form smooth and elongated shape, dice and roll to round shape.
5. Boil half pot of water. Add in rice balls, boil until floating and swell. Drain and set aside.
6. Boil chicken soup. Dip metal spoon with water, use it to put little bit of dace fish meat into soup, and boil until floating. Add in Chinese preserved sausage, Jinhua ham, coriander and rice balls. Add salt to tatse and serve.

肥牛鹹湯圓
Rice Balls in Beef Soup

數量：2～4人份 / serves
烹製時間：30分鐘 / mins

材料

湯料

肥牛 10 片
高湯 1000 克
菜心 2～3 棵

湯圓料

糯米粉 70 克
溫水 60 克

調味料

鹽 少許
胡椒粉 少許

做法

1. 菜心洗淨，瀝乾。
2. 糯米粉放在大碗中，慢慢加入溫水，搓揉成軟滑有光澤的長條，切粒，搓圓。
3. 燒滾半鍋清水，放入湯圓煮至浮起及脹身，撈出湯圓，備用。
4. 高湯燒滾，下鹽調味，放入菜心煮熟，加入湯圓，最後加入肥牛汆熟，下胡椒粉即可享用。

Ingredients

Soup

10 pcs thin sliced beef
1000g broth
2～3 choysum

Rice bal

70g glutinous rice flour
60g warm water

Seasonings

some salt
some pepper

Methods

1. Wash choysum and drain.
2. Put glutinous rice flour into large bowl. Add in warm water. Mix well to form a smooth dough. Roll to elongated shape. Dice and roll each to round shape.
3. Boil half pot of water, add in rice balls, boil until the rice balls floating and swell. Remove and set aside.
4. Bring broth to a boil, add salt to taste. Add choysum, cook until done, add cooked rice balls and thin sliced beef, cook briefly. Sprinkle with pepper. Serve in small bowls.

TIPS:

肥牛快熟，放入湯內汆幾十秒即可。
Thin sliced beef can be well-done in boiling soup for seconds. Don't be over-cooked.

PART 4
自家餡料變出新花樣

紫薯餡

材料

紫心番薯 400 克
砂糖 45 克
無鹽奶油 45 克

做法

1 紫心番薯沖洗乾淨，去皮、切塊，隔水蒸熟，取出，趁熱用湯匙壓成泥。
2 加入砂糖、無鹽奶油拌勻成糰，待完全涼透後包裹保鮮膜，放入冰箱存放。

TIPS:
番薯要趁熱才容易壓成泥，趁熱加入砂糖和奶油混合拌勻。

番薯餡

材料

黃番薯 400 克
砂糖 40 克
無鹽奶油 45 克

做法

1 黃番薯沖洗乾淨，去皮、切片，隔水蒸至熟，取出後趁熱用湯匙壓成泥。
2 加入砂糖、無鹽奶油拌勻成糰，待完全涼透後包裹保鮮膜，放入冰箱存放。

TIPS:
番薯要趁熱才容易壓成泥，趁熱加入砂糖和奶油混合拌勻。

花生醬餡

材料

粗粒花生醬 200 克
黃砂糖 60 克
糕粉（熟糯米粉） 適量

做法

1 粗粒花生醬加黃砂糖拌勻。
2 將糕粉逐次加入拌勻成糰，包裹保鮮膜，放入冰箱存放。

TIPS:
糕粉要逐次加入，太多會影響口感及味道，拌勻成糰即可。

豆沙餡

材料

紅豆 600 克
蔗糖 700 克
清水 1200 克

做法

1 紅豆沖洗後用清水浸泡 4 小時，濾掉水分，加入清水一起煮滾後，轉中小火煮約 1 小時，熄火燜 15 分鐘，再開小火熬煮約 1 小時至熟。
2 將紅豆沙放入炒菜鍋內，加入蔗糖用小火不停翻炒至糖溶解，再繼續炒至起黏狀可堆成山形。
3 盛起豆沙放涼，待完全涼透後包裹保鮮膜，放入冰箱存放。

TIPS:
◎豆沙餡冷卻後會收縮變硬一些，所以炒至有一點軟的程度就可以。
◎瓦斯爐各有不同火力，所以熬煮紅豆時需注意鍋內的水分，以免燒乾水。

黑芝麻泥

材料

黑芝麻粉 100 克
糖粉 50 克
固體豬油 適量

做法

1 黑芝麻粉加糖粉混合。
2 再加入豬油拌勻成糰，放入冰箱冰至凝固。

TIPS:
◎可以將花生放入鍋內炒香後搗碎，再加入芝麻粉拌勻，味道會更加香濃。
◎如果不想使用固體豬油，可以改用煮食油替代，而液態食油分量要自行調整，餡料不能太鬆散，否則難以成糰。

鳳梨餡

材料

新鮮鳳梨 1 個
黃砂糖 200 克
麥芽糖 100 克

做法

1 鳳梨去皮，切塊後刨成絲，鳳梨汁留著，備用。將鳳梨絲放入鍋內大火炒至有少許乾，加入部分鳳梨汁，每次收乾就再加入鳳梨汁直至鳳梨汁完全加入。（要不停翻炒以免黏鍋底）
2 當炒至軟熟及水分收乾呈糰狀時，加入黃砂糖，轉用中小火炒至乾。
3 最後加入麥芽糖炒至沒有湯汁，盛起，待完全涼透後放入冰箱存放。

TIPS:
◎如果在一開始炒鳳梨時就將全部汁液加入，會令炒餡時間增加。
◎鳳梨餡除了做湯圓餡，還可拿來做鳳梨酥或鳳梨卷。

紫薯紅莓餡

材料

紫薯餡 100 克
紅莓乾 30 克

做法

1 紅莓乾稍微切碎。
2 將紫薯與紅莓拌勻，再用力握緊成糰，即可。

TIPS:
可以將紫薯改為豆沙，味道也很合。

綠豆餡

材料

綠豆仁 300 克
砂糖 90 克
無鹽奶油 80 克

做法

1 綠豆仁沖洗後用清水浸泡1小時，瀝乾水分。
2 加入清水煮至熟。
3 趁熱將熟綠豆仁搗碎成綠豆泥。
4 將綠豆泥放入鍋內，炒至水分稍微蒸發，加入砂糖用小火不停翻炒至糖溶解，再加入無鹽奶油炒至成糰，盛起放涼，待完全涼透後放入冰箱存放。

TIPS:
◎煮綠豆仁的水一定要浸過豆面。
◎炒餡時可以加入少許桂花糖，便成桂花綠豆泥。

芋泥餡

材料

芋頭 600 克
砂糖 170 克
固體豬油 適量

做法

1 芋頭去皮、沖洗、切塊，隔水蒸熟，取出，趁熱用湯匙壓成泥。
2 加入砂糖並趁熱拌至糖溶解。若想餡料帶奶香味，可以加入一些奶粉增香。
3 再加入豬油拌勻成糰，盛起放涼，待完全涼透後放入冰箱存放。

TIPS:
芋頭的皮層會分泌黏液，皮膚觸及會產生劇癢，在處理時要戴上橡膠手套至削皮後再用水清洗。如果不小心被芋頭「咬」到，可以泡鹽水或用熱水洗刷一會兒。

檸檬糖漿

材料

砂糖 800 克
熱水 150 克
檸檬 3 ～ 4 片

做法

1 砂糖放入大容器內，將煮至 100℃的熱水以順時針方向逐次倒入，一邊倒入一邊手用長木湯匙攪鬆砂糖，要攪至糖全部溶解。
2 放涼後放入檸檬片，存放在冰箱，也可以加在飲料或甜品內。

蜜紅豆

材料

紅豆 300 克
清水 800 克
砂糖 220 克

做法

1 紅豆沖洗後用清水浸泡 4 ～ 6 小時（視天氣溫度而定），至紅豆體積膨脹至 2 ～ 3 倍，濾掉水分。加入清水煮滾，然後調小火煮約 35 ～ 45 分鐘，取出幾顆紅豆，試試夠軟爛即可。要注意紅豆皮不可破。
2 濾掉紅豆水，趁熱加入砂糖，用筷子小心地翻拌均勻，以免弄碎豆粒。如果攤涼後砂糖還沒溶解，可以連容器一起蒸或再放入熱水中，至糖溶解。放涼後放入冰箱存放。
3 蜜紅豆最少要提前一天製作，才能入味。

黑糖糖漿

材料

清水 100 克
薑 數片
黑糖 2 湯匙

做法

1 清水加入薑片一起煮滾，下黑糖，調小火熬煮成糖漿即可。